Kaiwan Jahanshahi

Crystallization of Polypeptides in Thin Films: Structures and Patterns

Kaiwan Jahanshahi

Crystallization of Polypeptides in Thin Films: Structures and Patterns

Südwestdeutscher Verlag für Hochschulschriften

Impressum / Imprint
Bibliografische Information der Deutschen Nationalbibliothek: Die Deutsche Nationalbibliothek verzeichnet diese Publikation in der Deutschen Nationalbibliografie; detaillierte bibliografische Daten sind im Internet über http://dnb.d-nb.de abrufbar.
Alle in diesem Buch genannten Marken und Produktnamen unterliegen warenzeichen-, marken- oder patentrechtlichem Schutz bzw. sind Warenzeichen oder eingetragene Warenzeichen der jeweiligen Inhaber. Die Wiedergabe von Marken, Produktnamen, Gebrauchsnamen, Handelsnamen, Warenbezeichnungen u.s.w. in diesem Werk berechtigt auch ohne besondere Kennzeichnung nicht zu der Annahme, dass solche Namen im Sinne der Warenzeichen- und Markenschutzgesetzgebung als frei zu betrachten wären und daher von jedermann benutzt werden dürften.

Bibliographic information published by the Deutsche Nationalbibliothek: The Deutsche Nationalbibliothek lists this publication in the Deutsche Nationalbibliografie; detailed bibliographic data are available in the Internet at http://dnb.d-nb.de.
Any brand names and product names mentioned in this book are subject to trademark, brand or patent protection and are trademarks or registered trademarks of their respective holders. The use of brand names, product names, common names, trade names, product descriptions etc. even without a particular marking in this works is in no way to be construed to mean that such names may be regarded as unrestricted in respect of trademark and brand protection legislation and could thus be used by anyone.

Coverbild / Cover image: www.ingimage.com

Verlag / Publisher:
Südwestdeutscher Verlag für Hochschulschriften
ist ein Imprint der / is a trademark of
OmniScriptum GmbH & Co. KG
Heinrich-Böcking-Str. 6-8, 66121 Saarbrücken, Deutschland / Germany
Email: info@svh-verlag.de

Herstellung: siehe letzte Seite /
Printed at: see last page
ISBN: 978-3-8381-3224-2

Zugl. / Approved by: Freiburg, Albert Ludwigs Universität, Diss., 2013

Copyright © 2014 OmniScriptum GmbH & Co. KG
Alle Rechte vorbehalten. / All rights reserved. Saarbrücken 2014

Abstract

Positional and orientational order of molecules has an important effect on the physical properties of the bulk materials. Therefore ordering of polymers in nano and microscale structures has been a hot topic in different fields of polymer science due to their application in organic electronics, bio inspired devices, etc. An effective and important way to achieve high order of polymer molecules on a large scale is polymer crystallization. Crystallization can be done either from polymer melts or polymer solutions. Crystallization of polymers from solution can lead to large microscale structures by efficiently controlling the parameters which promote and prevent nucleation, which is the first step in the crystallization process. The ability to have a control on these parameters provides a possibility for a more fundamental understanding of phenomena like ordering and growth, affected by competing intermolecular, molecule-solvent and molecule-nonsolvent interactions.

In this work we have investigated nucleation, growth and dissolution of poly(γ-benzyl L-glutamate) (PBLG) objects from semi-dilute thin film solutions of chloroform by condensation of methanol from the vapor phase onto the liquid film. The experimental approach which we adopted here, allows us to vary the quality of the solvent by condensation and evaporation of methanol in a controlled way by adjusting the saturation of methanol in the vapor phase and the temperature of the thin film solution. Adding and removing different amounts of methanol allowed us to reversibly control nucleation, growth and dissolution of objects possessing a hexagonal columnar liquid crystalline internal structure. Adding methanol to the isotropic polymer solution promoted nucleation and growth even at very low concentrated polymer solutions, i.e. significantly decreased the solubility limit (equilibrium volume fraction). Additionally, the variation of the number density of nuclei with the supersaturation ratio for various equilibrium concentrations was found to fit well the predictions of the classical nucleation theory. Based on our data and concentration regime that we have worked in, we conclude that at a specific supersaturation ratio the number of nucleated objects will be lower for lower equilibrium concentrations.

Abstract

After drying extraction of solvent and nonsolvent from thin film solution by fast evaporation, PBLG objects transformed to dry crystals. Each PBLG crystal possessed an internal domain structure exhibiting a zig-zag pattern consisting of parallel stripes of alternating orientations between domains. X-ray scattering and electron diffraction revealed a pseudo-hexagonal packing of the PBLG α-helices within these crystals with their axis oriented parallel to the plane of the substrate. Based on optical anisotropy studies, it could be shown that the orientation of the helix axis was parallel to the stripes. While forming in solution, the objects are assumed to consist of a hexagonal columnar liquid crystalline phase. Upon drying, lateral packing density of the helices increased and resulted in a net dilative strain perpendicular to the columns, which is supposed to cause the formation of zig-zag patterns.

Contents

Abstract   i

1 Introduction and Motivation   1
    1.1 Research Motivation 1
    1.2 Outline of Objectives and Research 3

2 Material and Experimental Techniques   5
    2.1 Material 5
       2.1.1 What are Polypeptides? 5
       2.1.2 Poly(γ-benzyl L-glutamate) 6
       2.1.3 Phase Diagram of PBLG in Solution 8
    2.2 Classical Nucleation Theory 11
    2.3 Experimental Methods 17
       2.3.1 Optical Microscopy 17
       2.3.2 Atomic Force Microscopy 19
       2.3.3 X-Ray Reflection 23
       2.3.4 Transmission Electron Microscopy 26
       2.3.5 Nuclear Magnetic Resonance (NMR) 28
       2.3.6 Ellipsometry 30

3 Systematic Control of Nucleation, Growth and Dissolution of PBLG Liquid Crystals in Thin Film Solutions   31
    3.1 Introduction: Controlling the Nucleation Density 31
    3.2 Sample Preparation 33
       3.2.1 Spin Coating 33
       3.2.2 Solvent Annealing Set Up 34
    3.3 Controlling the Volume Fractions 35
    3.4 Morphology and Growth Kinetics of the Liquid Crystalline Objects . 36

- 3.5 Reversible Nucleation and Dissolution of PBLG Liquid Crystals by Adding and Removing Methanol 38
- 3.6 Controlling Nucleation Rate, Number Density and Size of PBLG Liquid Crystals 40
- 3.7 NMR Investigation on PBLG Methanol Complexation 47
- 3.8 Conclusion 52

4 Structure, Pattern Formation and Orientation of PBLG Molecules within the Crystals 53
- 4.1 Introductory Remarks and State of the Art 53
- 4.2 Transformation from Single Domain to Multi Domain 54
- 4.3 Surface Topography Measured by AFM 57
- 4.4 Diffraction Measurements 58
  - 4.4.1 X-Ray Measurement 58
  - 4.4.2 Electron Diffraction Measurement 59
- 4.5 Birefringent Domains and Molecular Orientation 62
- 4.6 Zigzag Pattern Formation 66
- 4.7 Conclusion 68

5 General Conclusions and Perspectives 71

Bibliography 75

Appendix A 83

Appendix B 93

Acknowledgment 97

1 Introduction and Motivation

1.1 Research Motivation

Crystallization is a well-studied process that is used for many years to obtain well ordered structures. In this field, polymer crystallization which attracted many interests, refers to alignment of polymer molecules initiated from the melt, mechanical stretching or solution to form crystalline structures on various length scales [1–3]. Polymer crystallization has attracted considerable attention for its use in the design and fabrication of polymeric nanostructures leading to the design and development of advanced and new functional materials. Macromolecules with well-defined structures on the nanometer scale are perfect candidates for crystallizable materials and nanometer to micrometer scale devices [4].

The first polymers observed to form a liquid crystalline phase were of biological origin, particularly the rod like tobacco mosaic virus [5–8]. Many additional biological polymers are known to form ordered phases, including a number of globular polymers which reversibly polymerize to form long rod like molecules [8, 9]. The first [8, 10] synthetic polymer observed to form a liquid crystalline phase, poly(γ-benzyl L-glutamate), was studied before it was known to disperse as a rod like α-helix [8, 11].

Studies of PBLG have had an essential role not only in understanding the structure and solution properties of polypeptides but also in the basic understandings of polymer crystallization from solution . The α-helix model of Pauling was first verified by crystallographic studies on PBLG fibers by Perutz [12], and liquid crystallinity in a synthetic polymer solution was first observed by Robinson with PBLG [13]. Since the early studies, PBLG was found as a most useful model system for rigid-rod polymers [8].

The PBLG rods are rendered soluble by the mixing of the flexible side groups with the solvent. PBLG molecules can aggregate in some solvents, in particular close to conditions where gelation sets in. The presence of benzyl side groups together with polar end groups in the rods can help to form such aggregations even in dilute solution [14]. At sufficiently high concentrations (for example 130 mg/ml to 220 mg/ml for DNA molecules dissolved in water [15]), PBLG solutions form a cholesteric liquid crystalline phase with a relatively long pitch (up to 100 μm), as evidenced by the characteristic "fingerprint" pattern observed by optical microscopy between crossed polarizers [13]. A scattering peak, due to the inter-rod spacing in the cholesteric phase, is observed in SAXS measurements, and the peak position shifts to smaller spacings as the polymer concentration increases [14, 16]. The phase diagram of PBLG in DMF exhibits a narrow and a wide biphasic regions [17] in qualitatively good agreement with Flory's theory [18, 19]. At even higher concentrations (e.g., above about 20% in DMF) formation of a hexagonal columnar phase, characterized by sharp x-ray reflections, has been reported [14, 16]. The crystal structure of this phase, in which the unit cell is larger than that of the solid PBLG fiber, has been determined [14, 20, 21]. It involves hexagonal packing of single helices enjoined by a complex stacking of benzene rings from the side groups of neighboring helices which then results in a pseudohexagonal structure [14,22]. Russo and Miller [23] attributed its formation to the presence of water or methanol, a non-solvent, in the PBLG / DMF system.

Different phases which have been observed in PBLG solutions strongly depend on the concentration of the solution and the nature and power of the solvent [14, 24, 25]. Controlling the solvent power in solutions of PBLG or rod-like polymers in general has huge influence on their phase behavior. it is an important way to induce a separation of an isotropic solution into two coexisting phases, where a higher concentrated anisotropic, liquid crystalline phase is formed, coexisting with a lower concentrated isotropic solution.

The first step in phase transitions and in the presence of a thermodynamic barrier starts with nucleation, an important process that controls the number of objects of the newly formed liquid crystalline aggregates within the isotropic phase. Therefore, it is important to have a precise control on such phase transitions in order to study properties and behavior of each phase. In dilute solution nucleation process can be initiated by either increasing the polymer concentration [2] or decreasing the solubility limit of polymers [3,4]. So, it is important to have control on both polymer concentration and solubility limit of polymer solution in order to study the phase behavior of polymers. An efficient way to control the solubility limit (equilibrium volume fraction) of a polymer solution is to change the power of the solvent by temperature variation [2]. Another novel way of changing the solubility limit without changing the temperature which is presented in this thesis requires adding controlled amounts of nonsolvent to the solution which allows to initiate phase transition even at very low polymer concentrations. It will be shown in this thesis that this approach has important advantage of controlling the process of nucleation and growth of columnar hexagonal liquid crystalline phases in PBLG dilute solutions.

1.2 Outline of Objectives and Research

The objective of this thesis is to study and control the process of nucleation and growth of PBLG objects close to the transition from an isotropic solution to an anisotropic liquid crystalline phase in a reversible manner. The state and quality of the solution prior to the transition is an important condition. Solutions of PBLG molecules can exhibit a wide variety of mesophases and textures which may influence the formation of structure. In this thesis, the following case is considered: an isotropic solution and a hexagonal columnar liquid crystalline phase which transforms to a crystalline structure during the process of drying. The thesis is structured in the following way:

In chapter 2, the used materials and instruments are briefly introduced. The physical principles and basic theory of optical microscope (OM), atomic force microscopy (AFM) as the microscopes for observing and probing thin films are briefly described. Furthermore, the theory which governs and explains the results of this thesis is presented and discussed.

In chapter 3, we demonstrate an efficient strategy based on a new approach for reversibly controlling nucleation, growth and dissolution of large single domains of PBLG hexagonal columnar liquid crystals up to hundreds of micrometers in length from thin film solution. Adding and removing different amounts of the nonsolvent methanol, by regulating its vapor flow rate and adjusting the saturation of methanol in the vapor phase, allowed us to reversibly control two important parameters namely the interfacial tension and the super-saturation ratio in the thin film solution. This procedure, along with controlling polymer volume fraction in the thin film solution, allowed controlling the number density, size and growth rate of these liquid crystals. Additionally, we have shown that the variation of the number density of nuclei with the supersaturation ratio for various equilibrium concentrations fits well with the predictions of classical nucleation theory.

Upon drying the grown single domain PBLG objects, they transform to birefringent crystals having an internal domain structure exhibiting a zig-zag pattern consisting of parallel stripes of alternating orientations between domains. In Chapter 4, the results of X-ray and electron diffraction measurements on these crystals to identify their internal lattice structure are reported. The crystals were found to possess the structure of a pseudohexagonal lattice. Using optical birefringence, it could be shown that the orientation of the PBLG helix axis was parallel to the stripes. During the process of drying the lateral packing density of the helices increased and resulted in a net dilative strain perpendicular to the columns, which is supposed to cause the formation of zigzag patterns.

Chapter 5 provides a brief summary of results and an outlook.

2 Material and Experimental Techniques

2.1 Material

2.1.1 What are Polypeptides?

Multiple amino acids can be bonded together by peptide bonds between the carboxyl and amino groups of adjacent amino acid residues in order to form a peptide. Each of these bonds is formed by the dehydration of the carboxyl group of one amino acid and the amino group of the next amino acid. A polypeptide then is a single linear chain formed by long continuous bonding between such peptides. The number of peptide bonds depends on how long the polypeptide chain is. One end of every polypeptide has a free amino group and is called the amino terminal or N-terminal. The other end, with its free carboxyl group, is called the carboxyl terminal or C-terminal (see Fig. 2.1). Additional information about polypeptides can be found in the literature [26–28].

The amino acids are defined by an amino group (NH_2) and a carboxyl group (COOH) connected to an alpha carbon (C_α) to which a hydrogen and a side chain group R are attached. The smallest amino acid, glycine, has a hydrogen atom in place of a side chain R. All other amino acids have distinctive R groups (in case of glutamic acid, the R group is given by $-CH_2-CH_2-COOH$). Because the C_α of the other amino acids have four different constituents, the C_α atom is an asymmetric center (chiral) and amino acids occur in two optical isomers: D (Dextrorotatory) and L (Levorotatory) forms. These forms represent the ways in which the amino acid spiral is wound up. The D form is the right wound type, while the L form is the mirror left winding amino acid. Amino acids can also occur in DL configuration with a mixture of D and L forms. Most naturally occurring amino acids are in the L form (for example poly L-glutamic acid). Amino acids fall into several naturally occurring

Material and Experimental Techniques

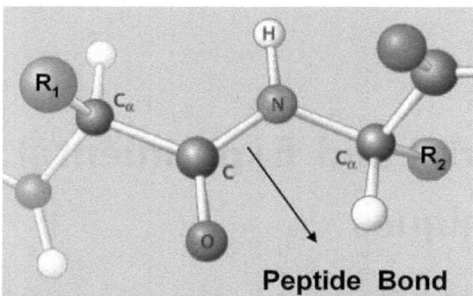

Figure 2.1: Illustration of the planar peptide bond formed between the carboxyl group of one amino acid and the amino group of another. The alpha carbon (C_α) atoms on either side of the peptide bond, the hydrogen atom (H) of the amide and the carbonyl oxygen atom (O) all lie within a plane. N symbolizes in the figure the nitrogen atom, C symbolizes the carbon atom, while R_1 and R_2 denote distinctive side groups. This figure adapted from [28]

groups including hydrophobic, hydrophilic, charged, basic, acidic etc. Glutamic acid ($C_5H_9NO_4$) is a hydrophilic amino acid [28].

Monosodium glutamate ($C_5H_8NNaO_4$) is a sodium salt of glutamic acid known in food industry as a food additive. In its pure form, it appears as a white crystalline powder. When dissolved in water it rapidly dissociates into free sodium and glutamate ions. Thus, glutamate is the anionic form of glutamic acid [28].

If an apolar benzyl group is added to a polar L-glutamate then γ-benzyl L-glutamate is obtained. This compound is not anymore soluble in protic solvents. As we will see at the end of this work, such insolubility can be of major interest in ordering of polypeptide based homopolymers at large scales.

2.1.2 Poly(γ-benzyl L-glutamate)

The chemical structure of poly(γ-benzyl L-glutamate) (PBLG), which was used in this thesis, is shown in Fig. 2.2a. Similar to natural polypeptides, synthetic polypeptides like PBLG can adopt an α-helical conformation [29] in the solid state [30] as well as in helicogenic solvents [31, 32]. The polymer backbone, built up by the amide groups, is stabilized by intra-molecular hydrogen bonds [29] between every main chain C=O and N-H group to a peptide bond 4 residues away yielding a regular, stable arrangement making an α-helical homo polypeptide with a pitch based on 18 monomer units in 5 turns. When formed by the residues of L amino acids (as polypeptides in nature), the helix is right-handed [33].

Polypeptide molecules of α-helical conformation can be regarded as stiff, rod-

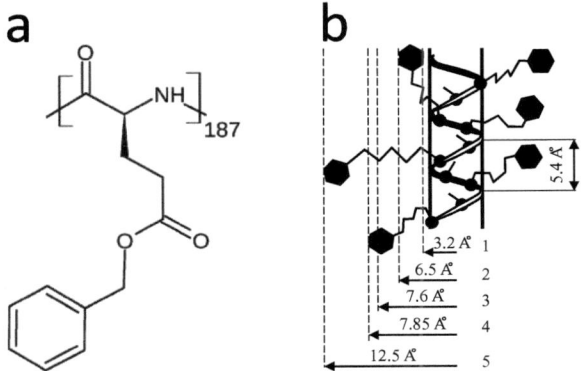

Figure 2.2: (a) Chemical structure of PBLG and (b) a schematic representation of the α-helical conformation of PBLG with its possible lateral dimensions. From top to bottom: 1- Radius of the backbone helix plus the first side bond length. 2- Half of the minimum intermolecular distance in a pseudo-hexagonal crystal lattice. 3- Backbone rod radius plus side chains represented as free rotators with root mean square dimensions. 4- Half of the intermolecular distance in crystallosolvate (for more information about crystallosolvates see [34]). 5- Backbone rod radius plus full extended side chains [34,35]. Fig. 2.2b adapted from [34].

like particles. The peptide planes are roughly parallel with the helix axis and the dipoles within the helix are aligned. All C=O groups point in the same direction and all N-H groups point into the opposite direction. The side chains point outward from the helix axis and are generally oriented towards the N-terminus [36]. The outer diameter is mainly determined by the conformation of these side groups (see Fig. 2.2b). In dry solid states, most frequently an average diameter of about 1.5 nm was found experimentally [21,34,37]. The largest packing distance for the hexagonal columnar phase of PBLG in m-cresol solution was found to be 1.84 nm [22], where the side-chains were swollen by the solvent. A hypothetical state with fully extended side-chains in trans-conformation [34,35] would correspond to a maximum diameter of 2.5 nm. The distance between monomers in the direction of the helical axis is

0.15 nm, resulting in 0.54 nm for one complete helical pitch [21, 29]. Although there are other structures possible [28, 36], the α-helix is the most common due to its stability and packing properties. Chirality, rigidity, geometrical anisotropy, hydrogen bonding [29] and a huge overall dipole moment caused by summation of the longitudinal components of the monomeric dipole moments between the N-H and C=O groups (3.5 Debye per monomer) [38] are key features of PBLG which have important effects on bulk properties of this polymer. Therefore, a variety of morphologies is anticipated to be found in thin film solutions at multiple lenghtscales starting from nanometer scale up to hundreds of micrometer scale. We aim using this system to control nucleation and growth process of PBLG liquid crystalline and crystalline structures up to large length scales which will give us a very deep understanding of this phenomena by linking it to the various physical parameters and intermolecular forces acting between the molecules.

We have summarized some of the most important specifications of PBLG in Appendix A.

2.1.3 Phase Diagram of PBLG in Solution

Construction of the phase diagram (PD) of polymer solutions capable of forming a liquid crystal (LC) state is an important direction in the investigation of the poly(γ-benzyl L-glutamate) / dimethylformamide (PBLG / DMF) system which is the most studied system of the rod like polymer-solvent systems [34, 39]. Unfortunately, the PD of PBLG / Chloroform system, used in this thesis, is not extensively studied. However, both DMF and chloroform are good solvents for PBLG hence, one can assume that the phase diagram of PBLG / Chloroform system is very similar to PBLG / DMF system. Construction of such a PD is due to both scientific interest in the polymer LC state, and of the application significance of the corresponding results. Thus, from LC polymer solutions high strength and high modulus fibers have been obtained [34, 40]; from polymer LC systems with polymerizable solvent anisotropic plastics have been obtained saving the LC structure of the initial solution [34, 41]; LC polymer solutions have been used for preparation of highly selective membranes [34, 42], including membranes from block copolymers with mesogenic blocks [34, 43]; many biological systems operate in the LC solution state [34, 44], etc. This is, naturally, far from the complete list of possible applications, but independently of its length, knowledge of the PD of the system used is undeniably necessary in all cases for its employment.

2.1.3.1 Two Component Systems

The phase behavior of binary solutions of stiff-chain polymers is quite different from the behavior of random coil polymers in that ordered or liquid crystalline phases are possible [45]. In this part we will present the PD of the bicomponent system of a rigid-chain polymer (PBLG) / isotropic solvent (DMF) that can form different phases in solution. Fig. 2.3 shows the most complete PD of the PBLG / DMF system. This shape of the PD is typical of PBLG with sufficiently high molecular mass 2×10^5-3×10^5 [34]. According to the experimental data [34,40–44,46,47] and in

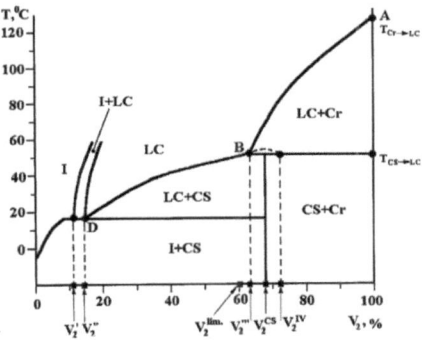

Figure 2.3: Schematic phase diagram of PBLG / DMF systems. Generalized PD, constructed on the basis of experimental literature data for PBLG molecular mass of $(2-3) \times 10^5$. We note that, by taking into account the molecular mass some changes is expectable in this generalized PD [34]. Here V_2 is the volume fraction of PBLG in solution.

full accordance with the concept of combination of various types of phase equilibria, the generalized PD of the system under consideration exhibits two homogeneity regions, corresponding to the isotropic liquid (I) and liquid crystal (LC), and five two-phase regions: isotropic liquid plus liquid crystal (I + LC), isotropic liquid plus crystallosolvate (I + CS), liquid crystal plus crystallosolvate (LC + CS), crystal plus crystallosolvate (Cr + CS), and crystal plus liquid crystal (Cr + LC) [34].

2.1.3.2 Three Component Systems

The phase relationships for a ternary system containing a rod like polymer, a solvent, and a nonsolvent has overall a similar shape as binary polymer / solvent system.

According to the theory developed by Flory [18] free energy of mixing ΔG for a system consisting of n_1 solvent molecules, n_2 nonsolvent molecules and n_3 rod like solute molecules is strongly depended on the thermodynamic interaction parameters χ_{i-j} between components of the mixture as well as other important parameters like temperature T, volume fraction of each component V_i and etc. Changing each of these parameters would lead to changes in the chemical potentials of the components in the mixture and will affect the position of the borders between isotropic and anisotropic phases in the mixture [48].

For rod like molecules in good solvent, the interaction parameter χ_{s-p} is almost zero (it can change a little with temperature) [23, 39]. For miscible solvent and nonsolvents the interaction parameter χ_{s-ns} between them is also zero. However the interaction parameter χ_{p-ns} between polymer and nonsolvent is nonzero and has a positive value. Positive values of χ_{p-ns} can cause phase separation even at very low polymer concentrations (depending on the value of χ_{p-ns} and volume fraction of nonsolvent).

Fig. 2.4 is the triangular representation of Nakajima's experimental results for a DMF / methanol / PBLG system with the volume fractions, V_1, V_2 and V_3 respectively.

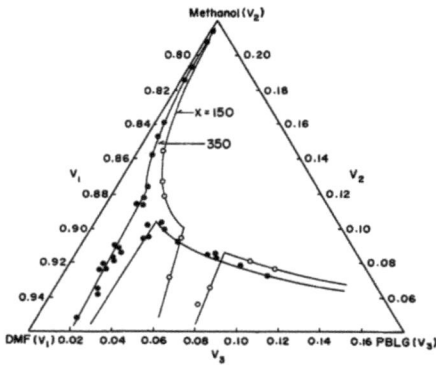

Figure 2.4: Phase diagrams at 30°C for systems composed of DMF, methanol, and PBLG with two different axial ratios: (o) x = 150; (•) x = 350 [39, 48].

The results obtained by Nakajama et al. [48] on experimental construction of the phase diagram for solution of PBLG (x = 150 and 350) in dimethylformamide may serve as an experimental corroboration of such a transition from a narrow to a wide region of coexistence of two equilibrium phases. Increasing the value of interaction parameter $\chi_{\text{p-ns}}$ by introducing methanol into the system, they initiated the transition from a narrow to a wide heterophase region for the methanol content of 0.10 to 0.12 volume fractions [39].

Phase transition from an isotropic solution to a new anisotropic phase starts by nucleating the new phase within the old phase which are separated from each other by an interface. The laws which govern such transitions are explained in the next part.

2.2 Classical Nucleation Theory

Nucleation can be promoted from solution either by cooling the solution, which leads to a decrease in solubility, or by evaporation. Crystal growth requires that nucleation should first occur and this can be achieved by one of a number of processes depending on the nature of the solution being examined [49]. Classical Nucleation Theory (CNT) is the simplest and most widely used theory that describes the nucleation process. Even though CNT was originally derived for condensation of a vapor into a liquid, it has also been employed "by analogy" to explain precipitation of crystals from supersaturated solutions and melts [50]. The thermodynamic description of this process was developed at the end of the 19th century by Gibbs, who defined the free energy change required for cluster formation ($\triangle G$) as sum of the free energy change for the phase transformation (bulk term, $\triangle G_{\text{v}}$) and the free energy change for the formation of a surface (surface term, $\triangle G_{\text{s}}$) at constant pressure and temperature. Nucleation from solution is described by the spontaneous tendency of a supersaturated solution to undergo precipitation and is equal to the number n_{tot} of units or molecules contributing to the nucleus multiplied by the difference between chemical potential of newly formed bulk phase μ_{b} at the current temperature and pressure of the surrounding metastable liquid phase and chemical potential of its surrounding liquid phase μ_l:

$$\triangle G = n_{\text{tot}} \triangle \mu = n_{\text{tot}} \left(\mu_{\text{b}} - \mu_l \right) \tag{2.1}$$

As the chemical potential of n_{s} of the molecules incorporating at the interface (μ_{s}) is different from chemical potential of the bulk molecules, we need to add correction terms to 2.1:

$$\begin{aligned}\Delta G &= n_{\text{tot}}(\mu_b - \mu_l) - n_s(\mu_b - \mu_l) + n_s(\mu_s - \mu_l) \\ &= (n_{\text{tot}} - n_s)(\mu_b - \mu_l) + n_s(\mu_s - \mu_l) \\ &= n(\mu_b - \mu_l) + n_s(\mu_s - \mu_l) = \Delta G_v + \Delta G_s\end{aligned} \quad (2.2)$$

Here n is the number of the molecules inside the bulk phase (not at the surface). Since the bulk phase is more stable than the liquid phase $\mu_b < \mu_l$, then ΔG_v becomes negative which would decrease the Gibbs free energy of the system. Also, in supersaturated solutions, this term can be expressed as [51]:

$$\Delta G_v = -n\,kT \ln S \quad (2.3)$$

Here k is Boltzmann constant, T is absolute temperature and S is the supersaturation ratio $\dfrac{\varphi_p}{\varphi_e}$, where φ_p is the volume fraction of the solute in the supersaturated solution and φ_e is its equilibrium volume fraction, i.e. the volume fraction of solute dissolved in a solvent, which is in equilibrium with the formed new phase. Thus, φ_e is the maximum amount of solute that can be dissolved in a given volume of solvent. Consequently, when $\varphi_p < \varphi_e$ no stable nuclei can form [51,52]. The second term (surface term) in 2.2 is the product of the surface area A of the formed nuclei and the interfacial tension σ between nuclei and its surrounding liquid phase [51,52] which can be expressed in the following way:

$$\Delta G_s = n_s(\mu_s - \mu_l) = A\frac{(\mu_s - \mu_l)}{a_m} = A\,\sigma \quad (2.4)$$

Here a_m is the part of the molecular surface contributing to the interface. In this term, $\mu_s > \mu_l$ and make ΔG_s positive. It means that introduction of a solid-liquid interface increases the free energy by an amount equal to the surface area of the nuclei multiplied by the interfacial tension of the bulk and liquid phases. Now we can rewrite the Gibbs free energy in the following form:

$$\Delta G = \Delta G_v + \Delta G_s = -nkT \ln S + A\,\sigma \quad (2.5)$$

and it means that formation and growth of a nuclei in a supersaturated solution depends on the competition between a decrease in ΔG_v which promotes nucleation and an increase in ΔG_s which favors dissolution (see Fig. 2.5b).

2.2 Classical Nucleation Theory

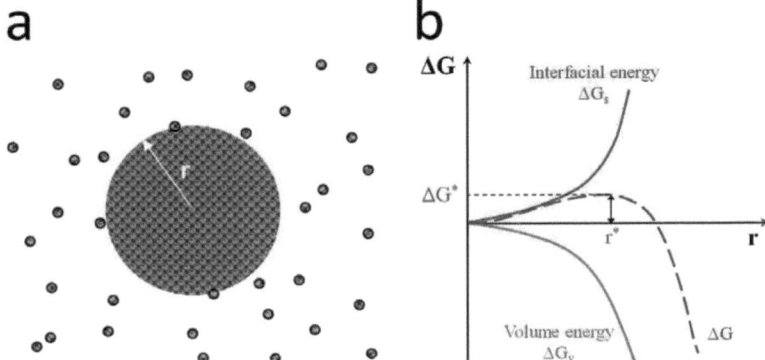

Figure 2.5: (a) Formation of a spherical nucleus of radius r from a solution leads to the free energy changes shown in b. (b) Schematic representation of Gibbs free energy vs. radius r of the nuclei. The cross-over of the bulk and surface terms combined with their opposing signs leads to a free energy barrier (ΔG^*) and a critical radius for nuclei. Figure adapted from [53, 54].

Volume V and Area A of the nucleus are assumed to be proportional to the third and second power of a generalized radius r of the nucleus (see Fig. 2.5a) and a geometry factor β defined by [51]:

$$V = \frac{\beta r^3}{4} \tag{2.6}$$

(corresponding to $\beta = 32$ for cubes and $\beta = 16\pi/3$ for spheres) and

$$A = \frac{3\beta r^2}{4} \tag{2.7}$$

With the volume contribution v per solute particle or molecule

$$n = \frac{V}{v} = \frac{\beta r^3}{4v} \tag{2.8}$$

then ΔG can be rewritten as

$$\Delta G = -\frac{\beta r^3}{4v} kT \ln\left(\frac{\varphi_p}{\varphi_e}\right) + \frac{3}{4}\sigma\beta r^2 \tag{2.9}$$

The first negative bulk term decreases as r^3, and the second positive surface term increases as r^2 and dominates at small radii, which causes an increase in the total free energy initially. Thus the smallest nucleus in solution typically dissolves. As the nuclei size increases the total free energy goes through a maximum denoted by $\triangle G^*$ at a critical size r^*, above which the total free energy decreases continuously and growth becomes energetically favorable, resulting in the formation of a stable nucleus. Taking the derivative of 2.9 with respect to r and equating to zero yields:

$$\left(\frac{\partial}{\partial r}\Delta G\right)_{r=r^*} = -\frac{3\beta r^2}{4v}kT \ln\left(\frac{\varphi_p}{\varphi_e}\right) + \frac{3}{2}\sigma\beta r = 0 \tag{2.10}$$

resulting in the critical radius

$$r^* = \frac{2\sigma v}{kT \ln(\varphi_p/\varphi_e)} \tag{2.11}$$

And the critical number of particles or molecules incorporating in the critical nucleus would be:

$$n^* = \frac{V^*}{v} = \frac{\beta r^{*3}}{4v} = \frac{\beta}{4v}\left[\frac{2\sigma v}{kT \ln(\varphi_p/\varphi_e)}\right]^3 = \frac{2\beta\sigma^3 v^2}{[kT \ln(\varphi_p/\varphi_e)]^3} \tag{2.12}$$

by inserting the equations for r^* and n^* in the 2.9 we obtain the critical Gibbs free energy which must be overcome in order to form a stable nucleus:

$$\begin{aligned}\Delta G^* &= -\frac{2\beta\sigma^3 v^2}{[kT \ln(\varphi_p/\varphi_e)]^2} + \frac{3}{4}\sigma\beta\left[\frac{2\sigma v}{kT \ln(\varphi_p/\varphi_e)}\right]^2 \\ &= \frac{\beta\sigma^3 v^2}{[kT \ln(\varphi_p/\varphi_e)]^2}\end{aligned} \tag{2.13}$$

In order to form a nucleus this energy barrier ($\triangle G^*$) must be exceeded. The probability to over come this or in the other words, the nucleation probability is proportional to:

$$P \propto \exp\left(\frac{-\Delta G^*}{kT}\right) \tag{2.14}$$

In the kinetic theory based on Gibbs formalism, the steady-state rate of nucleation J, which is equal to the number of formed nucleus per unit time and unit volume,

2.2 Classical Nucleation Theory

is expressed in the form of the Arrhenius reaction rate equation:

$$J = B \exp\left(\frac{-\Delta G^*}{kT}\right) = B \exp\left\{-\frac{\beta \sigma^3 v^2}{(kT)^3 \left[\ln\left(\varphi_p/\varphi_e\right)\right]^2}\right\} \quad (2.15)$$

The prefactor B is assumed to be only weakly depending on the volume fraction φ_p, resulting in

$$\begin{aligned}
\frac{d \ln J}{d \ln \varphi_p} &= \frac{d \ln J}{d \ln (\varphi_p/\varphi_e)} \approx \frac{d}{d \ln (\varphi_p/\varphi_e)}\left\{-\frac{\beta \sigma^3 v^2}{(kT)^3 \left[\ln(\varphi_p/\varphi_e)\right]^2}\right\} \\
&= \frac{2\beta \sigma^3 v^2}{[kT \ln(\varphi_p/\varphi_e)]^3} = n^*
\end{aligned} \quad (2.16)$$

Assuming a constant n^* results in

$$J \approx J_0 \left(\varphi_p/\varphi_e\right)^{n^*} \quad (2.17)$$

In the work of Mohanty [51] the dependence of the normalized profiles of φ_p, nucleation rate J and nucleation density N as a function of time (see Fig. 2.6) has been derived. Here N_∞ represents the number of nuclei per unit volume created up to the time t_1 (characteristic induction period at which point φ_p reaches the value of the equilibrium volume fraction φ_e). After t_1 nucleation is not possible anymore, it thus determines the total number of nuclei N_∞ which are experimentally observable (N_∞ can be determined by counting the structures on the film surface after drying the film, under OM or AFM) [28].

In order to compare our observations with theory we have to find a relation between the measured $N(t)$ and nucleation rate $J(t)$, how often nucleation happens per unit volume and unit time. The nucleation rate $J(t)$ is constant if there is an infinite reservoir in which there is no change in concentration. Under such conditions, nuclei are continuously appearing. In thin films of finite volume, we always have a decrease of the rate of nucleation (because the concentration of the solution film decreases). In Mohanty's work [51] it is assumed that all nuclei have been created before an induction time t_1. So, on average, after integration over the whole time during which nuclei were created, we obtain a value which does not change after t_1 anymore and which can be compared to our observations. Thus, N_∞ can be determined by integrating the nucleation rate J per unit volume of a homogeneous

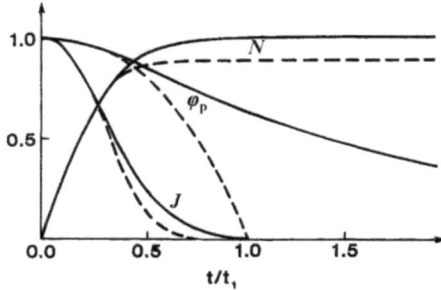

Figure 2.6: Normalized profiles of φ_p, J and N as function of time t, scaled with respect to the induction period t_1. The full curves correspond to the full-blown theory [51]; the broken curves result from the approximate expression for φ_p given in [51].

solution over time as shown in equation 2.18:

$$N_\infty \approx \int_0^{t_1} J \, dt \approx J_0 \int_0^{t_1} (\varphi_p/\varphi_e)^{n^*} \, dt \tag{2.18}$$

This is only valid when working at a low φ_p, very close to the φ_e, where very few nuclei are forming and which are not interacting. Once a nucleus is forming, it takes up so many molecules that around the nucleus the concentration drops below the φ_e. In addition, as there are many objects forming simultaneously which are then growing, they will deplete the surrounding region below φ_e. As the reservoir of available molecules will get exhausted the growth will stop. The nucleation density N which is the number of nuclei per unit volume after an induction period t_1 at which nucleation was stopped, is directly related to φ_e according to the following relation [51]:

$$N = P \exp\left\{-\frac{Q\beta\sigma^3 v^2}{(kT)^3 [\ln(\varphi_p/\varphi_e)]^2}\right\} \tag{2.19}$$

Here, Q is a dimensionless constant and P is a perfactor depending on the model used to describe growth after the induction period of nucleation [51]. Equations 2.15 for J and 2.19 for N show that supersaturation ratio S and interfacial tension σ are the main factors controlling nucleation kinetics. The nucleation rate increases when the supersaturation ratio increases and it decreases with increasing interfacial tension [55]. However, both S and σ depend on the equilibrium volume fraction φ_e. It is shown that σ depends linearly on the logarithm of φ_e [52, 55–57]. Correlation of these two parameters is discussed in Appendix B.

Finally, as demonstrated by Nielson and Söhnel [52], for a large variety of systems, σ depends on $\ln \varphi_e$, as expected from 2.5 for $\Delta G = 0$ [52]. By this assumption and also using the represented results in Appendix B and inserting B.15 in 2.19 and assuming spherical molecules ($\beta = \dfrac{16\pi}{3}$, $a_m = 2\pi r^2$ and $\nu = \dfrac{4\pi r^3}{3}$ i.e. neglecting the free space between close packed spheres) in a diffusion controlled process ($Q = \dfrac{3}{5}$), one obtains:

$$N = P \exp\left\{ -\frac{32}{45} \frac{[-\ln \varphi_e]^3}{[\ln \varphi_p - \ln \varphi_e]^2} \right\} \qquad (2.20)$$

This equation is used in this thesis in order to analyze the experimental data and extrapolate equilibrium volume fraction φ_e.

2.3 Experimental Methods

To achieve the goal of this thesis several investigation techniques were employed for characterizing the obtained crystals. In this part, we will discuss the experimental methods which were used in this work.

2.3.1 Optical Microscopy

An optical microscope (OM) is a type of microscope which uses visible light and a system of lenses to magnify images of fine objects which are not possible to be seen with the naked eye. An OM consists of a stand, a stage to hold the specimen, a movable body-tube containing the two lens systems, and mechanical controls for easy movement of the body and the specimen. The condenser lens focuses the light on the sample and the objective lens magnifies the beam, which contains the image, to the projector lens, so the image can be viewed by the observer. In microscopes, magnification is usually indicated by the abbreviation "X". The magnification power of the objective lens is the focal lengths of the tube lens divided by the focal length of the objective. To magnify an object even more, two lenses in combination have to be used (objective and eyepiece). The magnification of a microscope (when viewed through the eyepiece) is thus a product of $M_{\text{objective}} \times M_{\text{eyepiece}}$ [58, 59].

Thin polymeric films on silicon wafers show beautiful rainbow colors. The incoming light from the microscope is reflected at the polymer–air and at the polymer–silicon interface. The two reflected beams interfere with each other. If the differences in the refractive indices at the two interfaces are comparable, the reflected amplitudes will be on the same order of magnitude [60]. Therefore, it is possible that destructive interference can reduce the intensities of respective wavelengths considerably compared to wavelengths which interfere constructively. Thus, a selection of a single color is possible and is representative for a certain film thickness. The colors follow a periodic scheme with increasing film thickness, starting with a light brown (20 nm) up to an alternating light pink and green (1.5 µm)(see Fig. 2.7).

Figure 2.7: Interference colors from a polystyrene film on a silicon wafer which are exactly similar to that of PBLG films [28, 61, 62]. The film thickness changes from left to right (not on a linear scale): 20 nm (light brown), 70 nm (dark brown), 100 nm (dark blue), 140 nm (light blue) 200 nm (yellow), 250–280 nm (purple), 290 nm (blue), 310 nm (turquoise), 330 nm (green), 350 nm (yellow), 400 nm (light purple), 420 nm (green), 460 nm (yellow), 520 nm (pink) . . . alternating light green and pink up to approximately 1.5 µm until it changes to a transparent gray. Figure adapted from [63].

OM was used in this thesis to follow in real time and direct space the swelling of thin PBLG films during their exposure to chloroform and methanol vapors. We attempted to follow, in real time and direct space, formation of PBLG crystals and columnar hexagonal liquid crystals in thin film solutions. This was done under special conditions of polymer volume fraction φ_p close to the equilibrium volume fraction φ_e. Under such conditions, crystals / liquid crystals were big enough and sufficiently separated to be distinguished under the optical microscope. We also could estimate the size and density of PBLG crystals / liquid crystals in the thin film solution.

One of the techniques which can be combined with OM, to obtain more information about the molecular orientation in crystals, is polarized optical microscope (POM). POM technique is most commonly used on birefringent samples where the polarized

2.3 Experimental Methods

light interacts with the sample and generates contrast [64]. The sample is placed between two crossed polarizers (polarizer and analyzer). The intensity of light in the focal plane of the optical microscope is a function of the orientation of the optical axis of the crystal on the surface and the phase retardation of the incident polarized beam. Using the theory described by Dierking [64], the phase retardation of light transmission through the birefringent crystals is given by:

$$\delta = \frac{2\pi}{\lambda}(n_e - n_o)d \tag{2.21}$$

where, the extra-ordinary refractive index is given by:

$$n_e = \frac{n_\perp n_\parallel}{\sqrt{n_\parallel^2 \cos^2\phi + n_\perp^2 \sin^2\phi}} \tag{2.22}$$

the ordinary refractive index is given by:

$$n_o = n_\perp \tag{2.23}$$

where d is the thickness of the crystal, λ is the vacuum wavelength, and ϕ is the angle between the optical axis of the crystal and the projection of the incident wave vector of light. The transmitted intensity of light coming out of the analyzer is given by:

$$I = I_0 \sin^2 2\phi \sin^2 \frac{\delta}{2} \tag{2.24}$$

where I_0 is the light intensity after the polarizer, and is the ϕ azimuthal angle i.e. the angle between the analyzer and the projection of the optic axis onto the sample plane. From equation 2.24, it can be seen that the transmitted intensity depends on the orientation of the crystal molecules on the substrate [64]. In this thesis we used POM technique to evaluate the orientation of PBLG molecules within the crystal and with respect to the plane of substrate.

2.3.2 Atomic Force Microscopy

2.3.2.1 The Physical Principles

Atomic force microscopy has had a significant impact on physics, chemistry, biology and material science. Basic AFM modes measure the topography of a sample with the only requirement being that the deposited sample on a surface is rigid enough to withstand imaging.

In AFM, a sharp scanning probe collects local information. The AFM employs

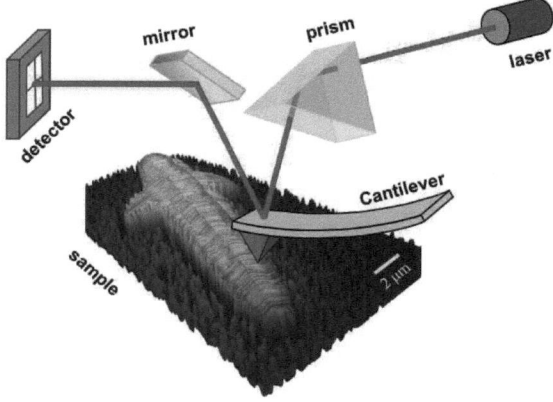

Figure 2.8: Schematic illustration of AFM. The tip is attached to a cantilever, and is raster-scanned over a surface. The cantilever deflection due to tip-surface interactions is monitored by a photodiode sensitive to laser light reflected at the tip backside. For microscopy applications, the position of the reflected beam is kept centered in the diode through feedback-controlled z-changes in the stage.

a microfabricated force-sensing flexible cantilever with small mass and a sharp tip to locally measure the attractive or repulsive forces acting between the surface of a sample and the atoms at (or near) the apex of the tip (see Fig. 2.8). These forces bend the cantilever towards or away from the surface, depending on whether the tip-sample interaction is attractive or repulsive, respectively. Optical detection systems and micro-fabricated cantilevers can detect forces in AFM down to the pico-Newton range. AFM can measure a variety of forces, including van der Waals forces, electrostatic forces, magnetic forces, adhesion forces and friction forces. Specialized modes of AFM can characterize the electrical, mechanical, and chemical properties of a sample in addition to its topography [65–67]. The cantilever bending is detected by laser beam deflection, as illustrated in Fig. 2.8. A laser beam is reflected from the back of the cantilever onto a split photodiode (most AFMs employ quadrant photodiode). Very small bending and motion of the cantilever is converted to a large

2.3 Experimental Methods

deflection of the laser spot on the photodiode which is then converted into voltage. The sum of the voltages; vertical deflection of the cantilever is determined by the difference between upper and lower halves of the photodiode, and lateral deflection is determined by difference between left and right halves of the photodiode. This voltage is compared with the set-point that is also represented a voltage value. The feedback loop regulates the movement of the piezoactuator at the cantilever base up or down to keep the cantilever deflection constant by adjusting the voltage applied to move the piezo. The 3D topography or height image of a surface is acquired from the magnitude of this voltage and also can be used to measure the tip-sample forces [59, 65, 66].

Depending on the application, and the required resolution there are different modes whether the tip is fixed and the sample is scanned or vice versa. The AFM measures the relative tip-sample displacement and any unwanted cantilever movement during scanning would add some vibrations. However, for large samples, AFMs are available where the tip is scanned over the sample [65–67]. The motion of the tip is enabled by a piezo-drive, which consists of three mutually perpendicular piezoelectric transducers: x piezo, y piezo, and z piezo (many AFMs use tubular piezo-electric tubes [65–67]). Upon applying a voltage, a piezoelectric transducer expands or contracts. By applying a sawtooth voltage on the x piezo and a voltage ramp on the y piezo, the tip scans on the xy plane. Using the coarse positioner and the z piezo, the tip and the sample can approach a few angstroms to each other.

2.3.2.2 AFM Operation Modes

The AFM can be operated in either static or dynamic modes [65, 66]. The static mode, is also called contact or repulsive mode. In this mode, the tip and the sample are brought into contact so that electronic orbitals of the atoms at the apex of the tip and the sample overlap (see Fig. 2.9a). Due to Pauli exclusion principle the tip experiences a very weak repulsive force. Since in this mode the tip never leaves the surface, this mode can be used for high resolution imaging such as atomic resolution. In contact mode, the lateral forces (the lateral deflection of the cantilever) between the tip and the sample can be used to measure friction forces.

In contact mode, the cantilever deflection is probed by a detection system. In this mode, deflections as small as $0.02\,nm$ and forces as low as $0.2\,nN$ can be measured. When the tip scans over the surface so that the height of the cantilever does not change, this is known as "constant height" imaging. However, it is much more

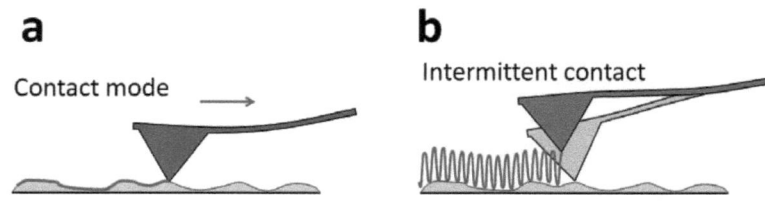

Figure 2.9: Schematic presentation of AFM operation modes. Figure adapted from [68]

common to use another type of imaging, the so-called "constant force" where a feedback loop is employed to monitor cantilever response to adjust the height of the cantilever in order to keep the deflection (force between the tip and the surface) constant, as the tip scans over the sample.

In the dynamic mode of operation, the tip is brought into very close distance from the sample (a few nanometers), but not into contact. In this technique, which avoids sample (or tip) crash or damage of the sample, the normal pressure exerted at the interface is close to zero. This makes the non-contact mode preferable to the contact mode. The cantilever is oscillated at a frequency slightly above or below its resonance frequency where the amplitude of oscillation is up to hundred nanometers (see Fig. 2.9b) [58,59,67,69,70]. The van der Waals forces, which are strongest from 1 nm to 10 nm above the surface, or any other long range force exerted above the surface acts to decrease the resonance frequency of the cantilever. This decrease in resonance frequency combined with the feedback loop system can be used to maintain a constant oscillation amplitude or frequency by adjusting the average tip-sample distance. Measuring the tip-sample distance at each (x, y) data point allows to construct a topographic image of the sample surface (see Fig. 2.10).

In the dynamic mode the cantilever vibration is either controlled by frequency modulation (FM) mode or more commonly, by amplitude modulation (AM) mode [67,71,72]. In frequency modulation mode, the information about tip-sample interactions, can be provided by changes in the oscillation frequency. Measuring frequency with very high sensitivity is possible by this technique and this allows for employing very stiff cantilevers and thus provides stability very close to the surface [72]. In phase modulation mode and based on stiffness variations associated with Young's

2.3 Experimental Methods

modulus change, AFM is able to record phase image of the sample [65, 66]. By mapping the phase of the cantilever oscillation during this mode of scan, a phase image can be recorded. Phase imaging helps to detect variations in composition, adhesion, friction, viscoelasticity, and perhaps other properties of the surface materials. Phase imaging results when there is a difference in phase between the imposed oscillation signal and the detected oscillation of the cantilever. This phase shift results from the dissipation of energy occurring during the tapping of the tip on the surface [65, 66]. Different materials will induce different energy dissipation, allowing their differentiation in an image, even on a topographically flat surface. Phase imaging is effective in the here studied thin polymer films especially for characterizing PBLG crystals.

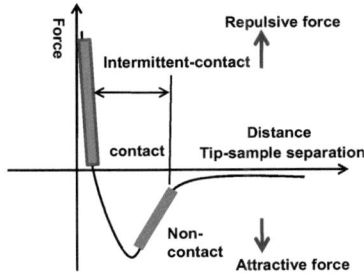

Figure 2.10: Inter-atomic forces as a function of probe - sample separation [58, 59].

In our study we have successfully employed AFM technique in ambient air with the aim to reveal the resulting ordered morphologies formed in thin films after their exposure to solvent vapor or the shape and surface roughness of polymer single crystals. AFM also provided valuable information on nucleation density of the ordered structures and also on the size and growth of polymer objects.

2.3.3 X-Ray Reflection

2.3.3.1 Principles of X-Ray

X-ray photons are a form of electromagnetic radiation produced following the ejection of an inner orbital electron and subsequent transition of atomic orbital electrons from states of high to low energy. When a monochromatic beam of X-ray photons falls onto a given specimen three basic phenomena may result, namely absorption, scattering or fluorescence. The coherently scattered photons may undergo subsequent interference leading in turn to the generation of diffraction maxima. These three basic phenomena form the bases of three important X-ray methods: the absorption technique, which is the basis of radiographic analysis; the scattering effect, which is the basis of X-ray diffraction; and the fluorescence effect, which is the basis of XRF spectrometry [73]. In our measurement we used the scattering effect in order to determine one of the lattice parameters of PBLG crystals.

The second field of materials analysis involves characterization by means of atomic or molecular arrangement in the crystal lattice. X-ray diffraction (XRD) uses single or multiphase specimens comprising a random orientation of small crystallites, each of the order of 1–50 μm in diameter. Each crystallite in turn is made up of a regular, ordered array of atoms or molecules. An ordered arrangement of atoms or molecules (the crystal lattice) contains planes of high atomic or molecular density which in turn means planes of high electron density. A monochromatic beam of X-ray photons will be scattered by these atomic or molecular electrons and if the scattered photons interfere with each other, diffraction maxima may occur. In general, one diffracted line will occur for each unique set of planes in the lattice. A diffraction pattern is typically in the form of a graph of diffraction angle (or interplanar spacing) against diffracted line intensity [73]. By using Bragg's law one can translate this specific diffraction angle to periodicity in the crystalline structure and determine the lattice parameters.

2.3.3.2 Lattice Parameter and Bragg's Law

Many materials are crystalline, that is their atoms or molecules are arranged in repeatable 3- dimensional arrays. These crystals are formed of unit cells, which contain the smallest number of atoms or molecules that repeat to form the 3-dimensional array. The unit cells can be cubic, hexagonal, or a few other types. The dimensions of these unit cells are called the lattice parameter. In the case of a cubic cell, only one lattice parameter is required to define its dimensions. One method of determining the lattice parameter of a crystalline solid is using Bragg's law with XRD, to measure the interplanar spacing [74, 75]. The interplanar spacing is the distance between two parallel planes of atoms or molecules in a crystalline material. The interplanar spacing can then be used to determine the lattice parameter. By knowing the coordinates of the diffracting plane, the intercept of the diffraction plane

2.3 Experimental Methods

with the x, y, and z axis, the lattice parameter can be determined. The correlation between the lattice parameter a_0 and the interplanar spacing d is defined in 2.25

$$d = \frac{a_0}{\sqrt{h^2 + k^2 + l^2}} \tag{2.25}$$

where h, k, and l are Miller indices or in fact the intercepts of the diffracting plane with the x, y and z axis respectively. Measuring the lattice parameter using Bragg's law involves two concepts [75, 76]. The first concept is interference of waves. When two waves come together, having the same wavelength, and frequency, the resultant waveform is the sum of the two waves. If the two waves of the same frequency are in phase, i.e. their amplitude maxima occur at the same time, the resultant amplitude will be the sum of the two amplitudes. If they are 180° out of phase, in other words, one amplitude maxima occurs exactly midway between the amplitude maxima of the other wave, or rather the peak of one wave occurs when the other wave at its trough, one observes destructive interference. If these two waves came from the same source, they would have the same initial amplitudes, and therefore the difference between amplitudes would be zero, that is the two waves would cancel each other out. The second concept of Bragg's Law involves simple trigonometry. When two waves hit atoms or molecules on two parallel lattice planes, one wave will travel an extra distance shown as d' in Fig. 2.11, and the same extra distance d' after diffracting off of the atoms, therefore one wave will travel $2d'$ greater distance than the other. The distance $2d'$ that one wave travels farther than the other is a function of the distance between the two planes (shown as d in Fig. 2.11), and the angle they make with the lattice plane [75].

From Fig. 2.11 it can be seen that d is the hypotenuse in a right triangle, and d' is the side opposite angle , therefore $d' = d \sin \theta$. If the extra distance that one wave travels $\left(2d'\right)$ is exactly equal to one full wavelength (λ) or any integer multiples of λ (i.e. $n\lambda$), then the waves will be back in phase again, and there will be a constructive interference. One could also say that this is a phase shift of 360°, which brings the two amplitude maxima back in phase. Since maximum in amplitude occurs only when the phase shift is equal to the wavelength λ or a multiple thereof $n\lambda$, the distance between two lattice planes (d) can be determined from the angle that the reflection occurs [75]. Equation 2.26 is Bragg's equation

$$d = \frac{n\lambda}{2 \sin \theta} \tag{2.26}$$

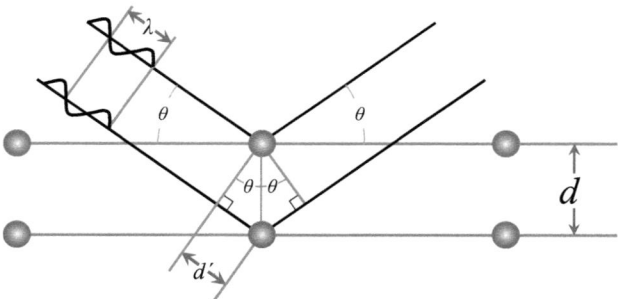

Figure 2.11: Determination of lattice spacing using Bragg's law. Figure adapted from [75]

which we have used in our experiments and in this thesis in order to determine one of the lattice parameters of PBLG crystals.

2.3.4 Transmission Electron Microscopy

Transmission electron microscopy (TEM) is a microscopy technique that uses a beam of electrons to examine samples on a very fine scale to gain information about morphology, composition and crystallographic properties. The electron beam interacts with the sample when it passes through the sample. There are three types of interactions of electron beam-specimen: (1) Un-scattered electrons (transmitted beam); (2) elastically diffracted electrons (diffracted beam) and (3) inelastically diffracted electrons.

The intensity of transmitted electron depends on the sample thickness. An image is formed from the interaction of the electrons transmitted through the specimen; the image is magnified and focused onto an imaging device e.g. CCD camera.

For crystalline structure, elastically diffracted electrons follow the Bragg's equation represented in 2.26 [77] where λ would be the wavelength of the electron beam, θ would be the angle between the incident beam and the surface of the crystal and d would be the spacing between layers of atoms or polymer molecules in our case. All electrons which are diffracted by the same atomic spacing will be diffracted by the same angle. These diffracted electrons are then collected using a magnetic lens and form spots on a screen. Each spot satisfied the diffraction condition reflecting the sample's crystal structure. The diffraction pattern gives information about the

2.3 Experimental Methods

crystal lattice parameters of the sample. A TEM apparatus is composed of (1) couple of condenser lenses to conduct the electron beam to the sample, (2) an objective lens to form the image/diffraction pattern of sample on the screen, (3) couple of intermediate lenses to magnify the image/diffraction pattern. (see Fig. 2.12).

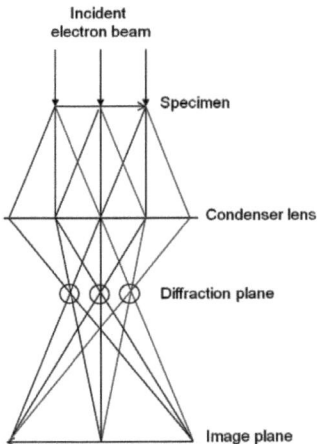

Figure 2.12: Schematic diagram showing the principles of the formation of SAED pattern and TEM image [78].

Moreover it is essential to understand the relationship between imaging and diffraction to determine the relationship between orientation of the crystal lattice and macroscopic features of a crystal. In order to obtain this information selected area diffraction (SAED) technique was developed. SAED is referred to as "selected" because the user can easily choose from which part of the specimen the diffraction pattern will be obtained. Located below the sample holder on the TEM column is a selected area aperture, which can be inserted into the beam path. This is a thin strip of metal that will block the beam. It contains several different sized holes, and can be moved by the user. The effect is to block all the electrons except for the small fraction passing through one of the holes; by moving the aperture hole to the section of the sample the user wishes to examine, this particular area is selected by the aperture, and only diffraction from the selected area will contribute to the SAED pattern on the screen [78]. In particular, this technique is useful to analyze single crystals. SAED pattern of a single crystal provides information about the symmetry of its crystal lattice and enables the calculation of d_{hkl} distances of Bravais lattice using the Bragg law [78]. Here, h, k and l are Miller indices of reciprocal lattice.

In order to determine the spacings in a crystalline lattice, the distances between spots on the diffraction image R should be measured. The d spacings can be found from

$$d = \frac{\lambda L}{R} \tag{2.27}$$

where R is the distance from the central bright spot to one of the rings or spots, L is the distance between sample and the diffraction plane and λ is the wavelength of the electron beam. The crystal lattice parameters in real space (a, b and c) as well as the angle between them (α, β and γ) can be derived using equation 2.28 [79]:

$$\frac{1}{d_{hkl}^2} = \frac{b^2c^2 \sin^2(\alpha h^2) + c^2a^2 \sin^2(\beta k^2) + a^2b^2 \sin^2(\gamma l^2) + 2abc^2 (\cos\alpha \cos\beta - \cos\gamma) h}{} $$
$$\frac{+2ab^2c (\cos\alpha \cos\gamma - \cos\beta) hl + 2a^2bc (\cos\beta \cos\gamma - \cos\alpha) kl}{a^2b^2c^2(1 - \cos^2\alpha - \cos^2\beta - \cos^2\gamma + 2\cos\alpha \cos\beta \cos\gamma)} \tag{2.28}$$

This technique was applied in this thesis to analyze the internal crystal structure of PBLG multi domain crystals and precisely determine the lattice parameters. Details are shown in chapter 4.

2.3.5 Nuclear Magnetic Resonance (NMR)

High-resolution NMR spectroscopy is a technique capable to provide detailed information on molecular structures based on atom nuclear interactions and properties. The theory of NMR was initially proposed by Pauli in 1924 who suggested that certain atomic nuclei should have the properties of spin and magnetic moment and that exposure to a magnetic field would consequently lead to the splitting of their energy levels (Pauli, 1924). However, it was first in 1946 that the NMR phenomena was experimentally discovered independently by Block & Packard (1946) and Purcell et al. (1946) and they were later awarded the Nobel price in physics 1952 [80].

Subatomic particles (electrons, protons and neutrons) can be considered as spinning around their axes. In atoms such as ^{12}C and ^{16}O, where the number of neutrons and protons are both even, these spins are paired, such that the nucleus of the atom

2.3 Experimental Methods

has no overall spin and cannot be detected by NMR. However, in some atoms, such as ^{1}H and ^{13}C, where the number of neutrons and/or the number of protons is odd, then the nucleus has a half-integer spin (i.e. $\frac{1}{2}, \frac{3}{2}, \frac{5}{2}$), and the nucleus does possess an overall spin measurable by NMR [80–82].

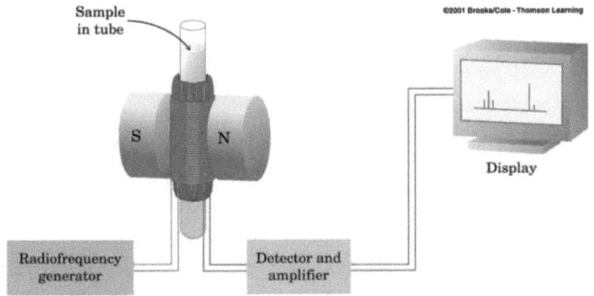

Continuous wave (CW) NMR
Pulsed (FT) NMR

Figure 2.13: Generalized schematic of NMR spectroscopy. Figure is adapted from [83]

NMR spectroscopy requires the application of strong magnetic fields and RF pulses to the nuclei of atoms (see Fig. 2.13). All nuclei are electrically charged, and those possessing a spin generate a small magnetic field. When an external magnetic field is applied, energy transfer is possible from the low-level to a high-energy level of the nuclei and their populations at different energy levels is governed by Boltzmann statistics. The energy transfer takes place at a frequency that corresponds to the radio frequency (RF), and when the spin returns to its low-level state, energy is emitted at the same frequency. The signal that matches this energy transfer is measured in several different ways and processed in order to give an NMR spectrum for the nucleus concerned. The precise resonant frequency of the energy transition depends on the effective magnetic field at the nucleus, and this field is affected by shielding of electrons orbiting the nucleus. Consequently, nuclei in different chemical environments absorb energy at slightly different resonance frequencies, and this effect is referred to as the chemical shift. This also means that sample conditions, such as pH and ion strength, will affect the observed spectrum. The chemical shift for ^{1}H NMR is determined as the difference in fractional units, δ (ppm), between the resonance frequency of the observed proton and that of a reference compound (In case of organic compounds generally tetramethylsilane, $(CH_3)_4\,Si$ (TMS) is used as a standard with respect to which chemical shift data is reported). The measured chemical shift of most protons is typically in the range of 0-10 ppm. A particular proton usually gives rise to more than one NMR signal because of the influence of non-equivalent neighboring protons, an effect called spin-spin coupling which is widely used in NMR technique in order to study the specific interactions of a particular proton with other compounds in the sample. The signal intensity depends on the number of identical nuclei, and is thus inherently quantitative [80–82, 84]. We applied this technique in this thesis, using a Bruker Avance II$^+$400 MHz NMR spectrometer, in order to study a possible complexation between solute PBLG molecules and methanol molecules. Details are shown and discussed in chapter 3.

2.3.6 Ellipsometry

Ellipsometry is a very precise technique to characterize thin films that measures the change in polarization of light upon reflection. The polarization change is represented as an amplitude ratio ψ and the phase difference $\triangle$. The measured response depends on the optical properties and thickness of the investigated materials. Thus, ellipsometry is widely used to determine film thickness and optical constants. However, it is also applied to characterize composition [85], crystallinity [86], roughness [87], doping concentration [88], and other material properties associated with a change in optical response [59, 89]. In this thesis, in order to determine the initial film thickness, ellipsometric measurements were made on PBLG thin films using an Optrel Multiskop ellipsometer with a 633 nm He − Ne laser illuminator.

3 Systematic Control of Nucleation, Growth and Dissolution of PBLG Liquid Crystals in Thin Film Solutions

3.1 Introduction: Controlling the Nucleation Density

In helicogenic solvents PBLG molecules can adopt an α-helical conformation [29, 31, 32] by stabilizing the polymer backbone via intra-molecular hydrogen bonds [29]. As a result, the molecules can be regarded as stiff, rod-like particles. For rods with a sufficiently high aspect ratio, when at the same time the volume fraction of the rods exceeds certain limits, liquid crystalline states can form in solutions. This has been predicted by the theories of Onsager [90] and Flory [18]. These theories were partially confirmed by experimental data [34, 39]. However, the system of PBLG in a helicogenic solvent can show a much more complex phase behavior, including the occurrence of different liquid crystalline states. For PBLG in dimethylformamide, the major findings concerned the existence of two homogeneous regions (isotropic and liquid crystalline) and five two-phase regions, where isotropic liquids, liquid crystalline phases, crystallosolvates or crystals can coexist [34, 39]. It should also be mentioned that under certain experimental conditions, e.g. when using a monodisperse polymer, layer-like ordering of smectic phases may be possible [91]. Moreover, using a non-linear optical method [92], a theoretically predicted [93] polar nematic phase was observed for PBLG solutions. Also hexagonal columnar liquid crystalline states, showing a solid-like hexagonal order in two dimensions and a liquid-like in the third [94], have been found in solutions with a high volume fraction of PBLG [22, 95]. Starting from such states in solutions of PBLG in m-cresol, the progressive removal of solvent was found to decrease the lateral packing distance continuously until a dry state was obtained [22].

It has been shown that the presence of methanol or other nonsolvents decreases effectively the solubility of PBLG homopolymers [23, 96] and also of rod-coil block copolymers with PBLG blocks [61, 96]. This is consistent with the Flory theory, where increased values of the Flory-Huggins parameter χ, a dimensionless measure of the interaction energy between solute and solvent, are predicted to decrease drastically the equilibrium volume fraction φ_e of the rods in the region of coexistent of isotropic and anisotropic phases [18, 23]. Therefore, at different polymer volume fractions φ_p, changing the parameter χ by regulating the amount of the nonsolvent can cause shifting between different phases in the solution. Within this complexity of the phases and phase diagrams, we focused only on the highly ordered phases for which we have adopted an experimental approach allowing to microscopically follow the nucleation and growth process in real time and in direct space.

In this chapter we present a study of the nucleation, growth and dissolution of ordered objects, hexagonal columnar liquid crystals from semi-dilute PBLG thin film solutions. It was convenient to induce nucleation by condensing methanol from the surrounding vapor phase onto a thin film of a chloroform solution. To prepare these solutions, we started from spin-coated thin polymeric films on a silicon substrate. Due to fast solvent evaporation in the course of spincoating, PBLG molecules were not able to form large and highly ordered structures but rather formed randomly oriented small aggregates which were strongly disordered. Increasing the mobility of PBLG molecules by exposing the dry thin films to vapor of chloroform transformed this spincoated thin PBLG film into a solution on a planar substrate [28]. The here adopted experimental approach allows us to vary the power of the mixed solvent by adding / removing methanol through the vapor phase by controlled condensation / evaporation. This was achieved by adjusting the vapor pressure of methanol in the surrounding vapor phase and by varying the temperature of the sample, i.e. the thin film solution. Nucleation and growth of the PBLG liquid crystals can be followed by optical microscopy in real time and direct space. Having a sensitive control on both PBLG volume fraction φ_p and the equilibrium volume fraction φ_e can lead to hexagonal columnar liquid single crystals of hundreds of micrometers length. The obtained nucleation density was analyzed on the base of the CNT. Based on our experimental approach of condensing a bad solvent (methanol) from the surrounding vapor phase, we were able to grow objects even from rather dilute polymer solutions as the presence of methanol significantly decreased the solubility limit and thus the equilibrium volume fraction φ_e of PBLG. Using the surrounding vapor phase as a reservoir allowed to reverse the process; i.e ordered objects could be re-dissolved. This was achieved by improving the power of the solvent through controlled evaporation of methanol from thin film solution by reducing the methanol flow rate and thus reducing the vapor pressure of the nonsolvent in the surrounding vapor phase.

3.2 Sample Preparation

3.2.1 Spin Coating

PBLG was purchased from Sigma-Aldrich with an average molecular weight $M_w = 41000$ g/mol and a polydispersity $M_w / M_n \approx 1.2$, corresponding to an average degree of polymerization of 187 (resulting in an average molecular length of 28 nm for the α-helical conformation). Solid thin films of PBLG (initial thickness $h_0 = 50 \pm 2$ nm) were obtained by spincoating from chloroform solution (0.6 weight %) onto hydrophilic silicon substrates previously cleaned using UV-ozone treatment.

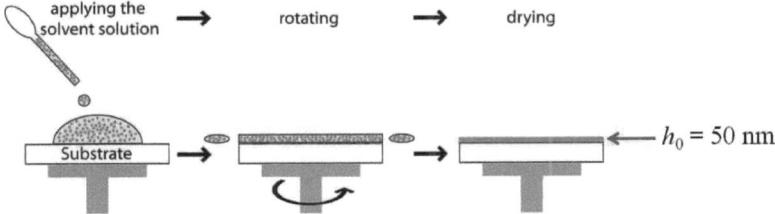

Figure 3.1: Schematic representation of the process of spin coating. Figure adapted from [97].

Silicon substrates were used to be able to determine the film thicknesses using an interference method described in previous chapter (see sec. 2.3.1). During spin coating, PBLG molecules spread randomly on the substrate and due to fast evaporation of the solvent, they do not have enough time to arrange themselves in large scale ordered structures. After spin coating and getting a dry thin film, the polymer molecules were not mobile because they were strongly interacting via non-specific and specific interactions like hydrogen bonds. However, to allow for ordering, a high mobility of molecules was the first step which had to be achieved in order to reach our goals. Thus, we needed to find methods to increase the mobility of the molecules. Exposing thin films to solvent vapor led to the formation of solutions with an increased mobility of the polymers by weakening the specific interactions and thus facilitated structure formation (the specific interactions could not be weakened by heating the polymers to high temperatures). That is why the influence of parameters like polymer volume fraction φ_p, the nonsolvent volume fraction φ_m, equilibrium volume fraction φ_e and interfacial tension σ on the process of controlling nucleation rate J, number density N and structure of PBLG liquid crystals and crystals was of interest [28].

3.2.2 Solvent Annealing Set Up

In order to study nucleation and growth of PBLG crystals in thin film solution, we annealed the spin coated PBLG thin films in a homebuilt chamber containing a peltier element and connected to two flow controllers (see Fig. 3.2).

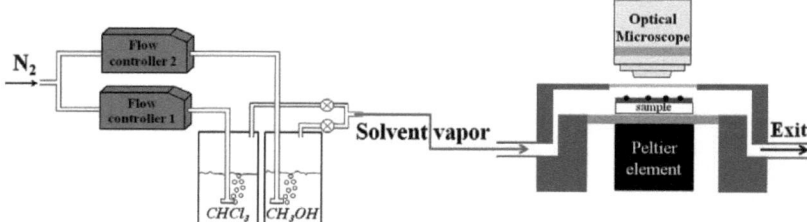

Figure 3.2: A schematic representation of the set up used to expose the samples to vapor phases.

In Fig. 3.2 a schematic of the setup used in my experiments is represented. The setup consists of two flow controllers connected to a nitrogen reservoir. The out coming nitrogen from the flow controllers passes through two bottles of solvent (chloroform) and nonsolvent (methanol) to achieve saturation of the gas. Afterwards, these two saturated nitrogen streams are mixed and then brought into the homebuilt chamber. The ratio of chloroform and methanol vapor present in the environment of the chamber can be controlled and regulated by the flow rates of the flow controllers. Before starting the experiment, the sample chamber must be flushed with pure nitrogen for about 5 minutes to remove any moisture or unwanted gas. The sample temperature is controlled by a peltier element located below the sample. Decreasing the sample temperature a few Kelvin below room temperature leads to condensation of solvent and nonsolvent onto the film and to swelling of it. This technically new approach allowed me to regulate sample temperature and vapor flow rates (ratio of solvent and nonsolvent) and hence, to determine the amount of solvent and non-solvent that condensed onto the sample. With this procedure two key parameters, the polymer volume fraction φ_p and the equilibrium volume fraction φ_e, can be controlled to initiate and observe the process of nucleation and growth of PBLG liquid crystals in real time and in a reversible manner (for more information see sec. 3.3). The whole growth process of the liquid crystalline objects was observed in situ through the window of the chamber using an optical microscope (Leitz, ORTHOLUX II POL-BK).

3.3 Controlling the Volume Fractions

As we already mentioned, I followed solvent and nonsolvent condensation in real time and direct space by using an optical microscope via the change of the interference colours of the film. As the amount of polymer in the film stayed constant (this quantity is proportional to the thickness of the dry spin-coated film), a change in film thickness was directly related to the amount of solvent and nonsolvent incorporated into the film, i.e. corresponded to swelling by solvent (see Fig. 3.3a). The volume fraction of PBLG was deduced from the ratio of the initial film thickness h_0 to the thickness of the swollen film $h = h_0 + h_c + h_m$, i.e., $\varphi_p = h_0 / h$. The volume fraction of methanol $\varphi_m = h_m / h$ was estimated by condensing, in a separate experiment where no chloroform solvent was present, only methanol on a dry PBLG film under similar temperature conditions (see Fig. 3.3b).

Here, h was derived from the interference colors shown in Fig. 2.7. For $h > 1600$ nm the number of interference fringes was counted. We note that the actual amount of the condensed methanol might be slightly different at conditions when chloroform is present too. Keeping the film at a constant and low temperature leads to a steady decrease of the φ_p and the φ_m due to continuously condensing solvent and nonsolvent molecules [28]. Thus, for experiments of long durations, the φ_p and the φ_m were kept roughly constant by continuously increasing slightly the sample temperature or regulating the flow rates of the solvent and nonsolvent vapors.

Finally, at a chosen time, the sample is being dried completely by simply heating the film to relatively high temperatures, for example to 65 °C. In summary, we

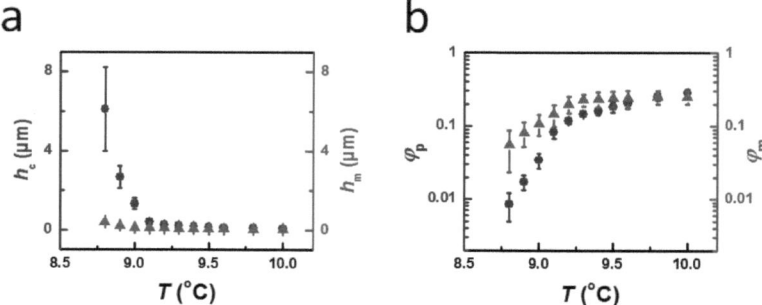

Figure 3.3: (a) Dependency of the film thickness contributions of condensed chloroform h_c and methanol h_m on the temperature. (b) PBLG volume fraction φ_p (blue circles) and methanol volume fraction, φ_m (red triangles) as a function of the film temperature.

distinguish three stages during nucleation and growth of PBLG objects from thin film solution:

1. condensing solvent onto the thin film and swelling it until low φ_p is reached.

2. adding controlled amounts of nonsolvent to the thin film solution in order to decrease φ_e and promote nucleation and growth of PBLG liquid crystalline objects.

3. complete drying of the thin film solution and getting dried PBLG crystals.

3.4 Morphology and Growth Kinetics of the Liquid Crystalline Objects

Nucleation of liquid crystalline objects in a supersaturated solution will be continued with the growth process because the rate of polymer molecules attaching to the surface of liquid crystalline object exceeds the rate of polymer molecules detaching from the surface. The kinetics of the attachment and detachment processes at step edges are determined by the energy barriers seen by the molecules. These barriers and bonds between polymer molecules and adjacent ones within the liquid crystalline objects would affect the growth process of the objects [54]. Polymer molecule itself

3.4 Morphology and Growth Kinetics of the Liquid Crystalline Objects

(conformation, length, side chains, etc.), its diffusion and concentration, limits the rate at which a liquid crystalline object can grow, often greatly affecting the final morphology of the object (see Fig. 3.4up).

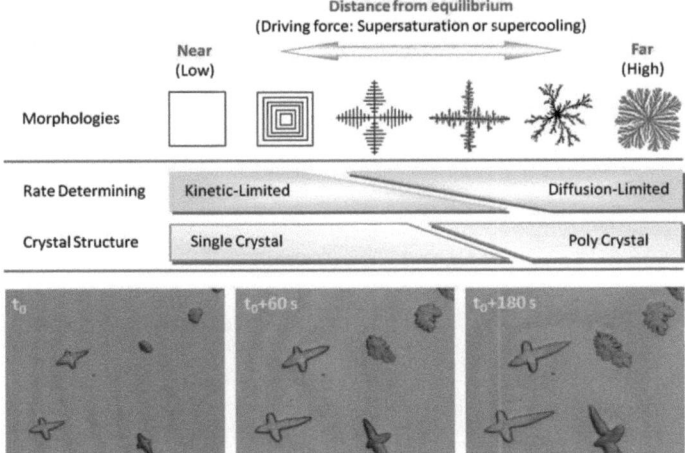

Figure 3.4: (up) Morphological variation of crystals with the changes of driving force [98]. (down) Series of optical micrographs showing growth process of PBLG objects with two different types of morphologies at the same experimental condition. The experiment was performed on a thin film of initial thickness $h_0 = 50$ nm swollen in chloroform until the polymer volume fraction reached $\varphi_p = 1.0 \pm 0.5\,\%$. Then nucleation was promoted by adding methanol $\varphi_m = 4 \pm 1\,\%$ to the system. All micrographs have a size of $188 \times 151\,\mu m^2$.

In Fig. 3.4down series of optical micrographs showing growth process of PBLG liquid crystalline objects are represented. As it is clear from micrographs, two types (cross like and dendritic) of PBLG objects can nucleate and grow to large sizes. The complicated kinetics of the attachment and detachment of PBLG molecules to the surface of nuclei and the number of growth fronts on the surface of each nuclei led to different morphologies. However, in this work we mainly focus on the nucleation process and controlling number densities rather than kinetic of growth and morphologies.

3.5 Reversible Nucleation and Dissolution of PBLG Liquid Crystals by Adding and Removing Methanol

Growing large scale PBLG liquid crystals by ordering molecules takes time, mainly due to the transport processes involved. During the spin coating process, the solvent was evaporating very quickly, i.e. within seconds, and consequently the molecules

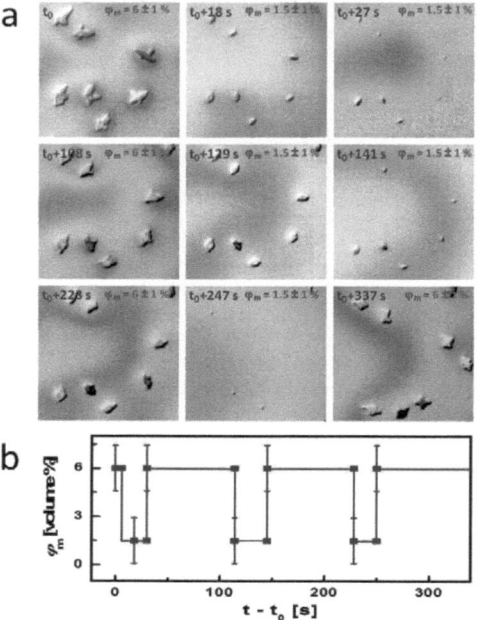

Figure 3.5: (a) Series of optical micrographs showing growth and dissolution of PBLG objects by adding and Removing of the methanol from the thin film solution. The experiment was performed on a thin film of initial thickness $h_0 = 50$ nm swollen in chloroform until the polymer volume fraction reached $\varphi_p = 2.5 \pm 0.5\,\%$. All micrographs have a size of $100 \times 100\,\mu\mathrm{m}^2$.
(b) Temporal evolution of methanol φ_m in the time course of the displayed micrographs.

3.5 Reversible Nucleation and Dissolution of PBLG Liquid Crystals by Adding and Removing Methanol

were randomly deposited onto the substrate and not organized on large scales. Some order may be present only at small scales. In the dry state, the polymers were not mobile because of strong non-specific and directional intermolecular forces. However, a sufficiently high mobility of molecules is necessary in order to allow for ordering. The here presented experiments, and additional ones for different concentrations which are not shown, illustrate that the results strongly depend on parameters like polymer volume fraction φ_p, the nonsolvent volume fraction φ_m, equilibrium volume fraction φ_e and interfacial tension σ.

In order to follow in real time the process of nucleation and growth, we have performed the experiments under the optical microscope (see sec. 3.2.2). The procedure consisted in swelling of a $h_0 = 50$ nm PBLG thin film in order to obtain high molecular mobility of polymers. We first cooled the film below room temperature and swelled it by condensing solvent vapor onto the film surface until getting an isotropic phase. Accordingly, φ_p decreased to about $\varphi_p = 2.5 \pm 0.5\,\%$ where the molecules were dispersed homogeneously and the molecular mobility was high. As φ_p was smaller than the equilibrium volume fraction φ_e, we added methanol $\varphi_m = 6 \pm 1\,\%$ to the thin film solution in order to decrease φ_e below φ_p in order to initiate nucleation and growth of PBLG liquid crystalline objects (see Fig. 3.5 at t_0). in the next section (see sec. 3.6) the effect of methanol on the equilibrium volume fraction will be discussed in more detail. After growing the PBLG objects, we decreased methanol saturation in the vapor phase above the thin film solution by decreasing it's flow rate. Hence, a part of methanol molecules evaporated and left the thin film solution in order to make equilibrium between two phases of methanol. Therefore, equilibrium volume fraction φ_e increased above the PBLG volume fraction φ_p and led to dissolution of the already formed PBLG objects (see Fig. 3.5 at $t_0 + 18$ s and $t_0 + 27$ s). In order to check the reversibility of the process and see if the objects can remember their previous morphology, we kept some small seeds i.e. before fully dissolution of objects we increased the amount of methanol $\varphi_m = 6 \pm 1\,\%$ in the solution. The remaining seeds start to grow again (see Fig. 3.5 at $t_0 + 108$ s) but with different shape; indicating that there is no memory of morphology for these objects. We repeated this process for three times and at each time these objects grew with different morphology (see Fig. 3.5 at $t_0 + 223$ s and $t_0 + 337$ s) which can be because of the complicated attachment of molecules to the growing surfaces of PBLG objects.

Beside the irreversibility of the morphology of the grown objects, it was promising that the nucleation process was reversible and could be controlled. the procedure of such a control is discussed in the next section.

3.6 Controlling Nucleation Rate, Number Density and Size of PBLG Liquid Crystals

After representing the reversibility of nucleation and growth of PBLG objects (see sec. 3.5), now we go one step further in order to control the process of nucleation

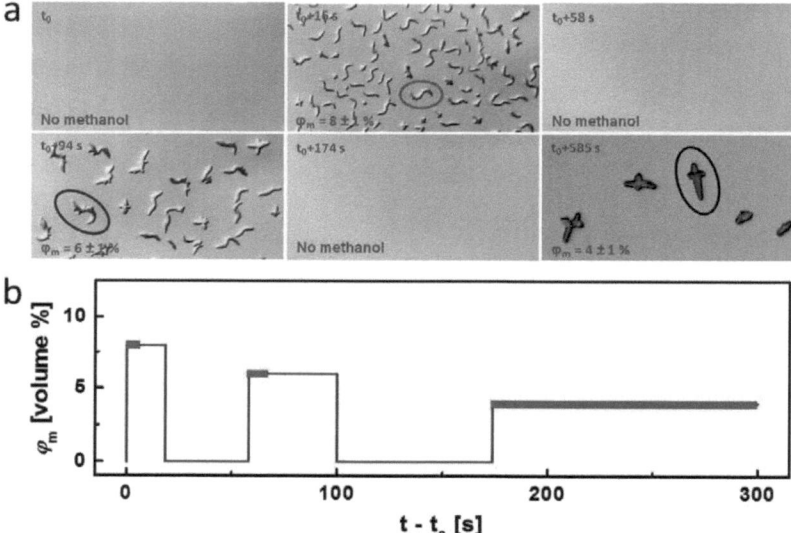

Figure 3.6: (a) Series of optical micrographs showing reversible growth and dissolution of PBLG crystals when adding or removing methanol from the thin film solution. The experiment was performed on a thin film of initial thickness $h_0 = 50$ nm swollen in chloroform until the polymer volume fraction reached $\varphi_p = 1.2 \pm 0.4\,\%$. All micrographs have a size of $230 \times 120\,\mu m^2$.
(b) Temporal evolution of methanol φ_m in the time course of the displayed micrographs. Thick red lines in the diagram represent the time interval one needs to wait after adding methanol to the thin film solution in order to observe the first nuclei formed.

3.6 Controlling Nucleation Rate, Number Density and Size of PBLG Liquid Crystals

and growth i.e. controlling including nucleation rate, number density and size of PBLG liquid crystalline objects (see Fig. 3.6).

Here again, we have performed the experiments under the optical microscope (see sec. 3.2.2) in order to control and follow in real time the nucleation and growth process. The procedure consisted in swelling of a $h_0 = 50\,\text{nm}$ PBLG thin film in order to obtain high molecular mobility of polymers. By cooling the film below room temperature solvent vapor condensed onto the film surface and made an isotropic phase. Accordingly, φ_p decreased to about $\varphi_\text{p} = 1.2 \pm 0.4\,\%$ where the molecules were dispersed homogeneously and the molecular mobility was high. This solution did not show any signs of birefringence between crossed polarizers. As the viscosity of the isotropic solution is comparatively low, surface tension was able to smoothen the surface of the film quickly within seconds (see Fig. 3.6a at t_0). This smoothening process and the absence of birefrengency between crossed polarizers is a clear indicator for the isotropic phase. As $\varphi_\text{p} = 1.2 \pm 0.4\,\%$ is smaller than the equilibrium volume fraction φ_e, no liquid crystalline objects were nucleated and we did not observe any changes in the thin film solution in time.

However, addition of methanol (volume fraction $\varphi_\text{m} = 8 \pm 1\,\%$) to this thin film solution at time t_0 led to nucleation and growth of many objects (Fig. 3.6a at $t_0 + 16\,\text{s}$). This can be explained by the fact that the presence of methanol decreases the equilibrium volume fraction φ_e below φ_p which in follow causes an increased interfacial tension σ between nuclei and the liquid surrounding phase [61, 96]. This leads to nucleation and growth of the liquid crystalline objects. Removing methanol from the film solution (by flushing pure nitrogen instead of nitrogen saturated with methanol) led to dissolution of all objects (Fig. 3.6a at $t_0 + 58\,\text{s}$). Adding again methanol (this time with $\varphi_\text{m} = 6 \pm 1\,\%$) led again to nucleation and growth of objects (Fig. 3.6a at $t_0 + 94\,\text{s}$). Their number density was lower, compared to the previous case when more methanol was present in the film solution. We did remove once more the methanol (Fig. 3.6a at $t_0 + 174\,\text{s}$) and observed the dissolution of all objects and then added once more even less methanol ($\varphi_\text{m} = 4 \pm 1\,\%$) leading to an even lower number density of objects (see Fig. 3.6a at $t_0 + 585\,\text{s}$). The experiments represented in Fig. 3.6 showed not only that nucleation, growth and dissolution of the objects is reversible but also allowed us to observe that adding more methanol to the film solution decreased the nucleation time (for example, the time interval between the moment we started to add methanol and first nuclei formation was about 6 seconds for 8 % of methanol, compared to about 2 minutes for 4 % of methanol).

Fig. 3.7 is emphasizing that the nucleation rate J as well as the crystal number density N is controllable by a precise variation of the volume fraction of methanol φ_m in the film solution. The quantities Jh and Nh presented in this figure are the products of J and N with the height h of the thin film solution, obtained by counting the number of nuclei per area of the optical micrograph and assuming that the nucleation density is homogenous throughout the thickness of the thin film

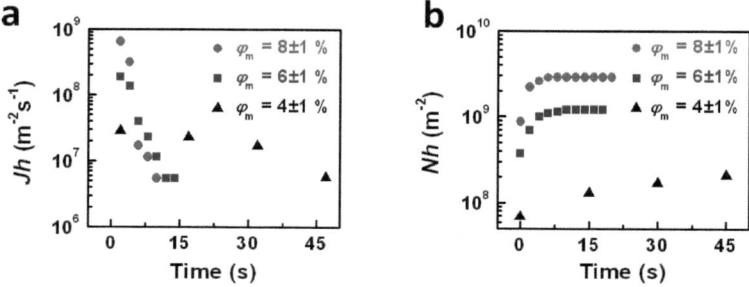

Figure 3.7: (a) Nucleation rate J and (b) number density N of nuclei multiplied by the thickness of thin film solution h which were observed in time when adding 8 %, 6 % and 4 % of methanol, respectively to the thin film solution with $\varphi_p = 1.2 \pm 0.4 \%$ (see Fig. 3.6).

solution. Both Jh and Nh increase with increasing φ_m (see Fig. 3.7a and Fig. 3.7b). For example, a thin film solution containing about 8 % methanol led to a final value of $Nh = 3.3 \times 10^9$ m^{-2} in comparison to a final value of 1.8×10^8 m^{-2} obtained for $\varphi_m = 4\%$. Note that Jh was found to decrease by about one order of magnitude for 8 % and 6 % of methanol in less than 15 seconds, while for 4 % of methanol Jh decreased much slower. Similarly, Nh saturated faster (less than 15 seconds) for 8 % and 6 % of φ_m compared to the case of $\varphi_m = 4\%$ where saturation occurred after about 60 seconds (not shown in Fig. 3.7b). These results can be explained by the fact that more methanol in the thin film solution lowers φ_e, i.e. increases S. Therefore, the probability of forming a nucleus is higher and needs less time (see 2.14). These experimental observations are in qualitative agreement with previous studies on star block copolymers [61, 96] and with predictions of the classical nucleation theory. In all cases Jh started with a maximal value, and then decreases in time to very low values. For the low values of $\varphi_p \approx 1\%$, one can expect values of φ_e still larger than that for the maximum in N at $\varphi_p^3/\varphi_e'^2$ (see Fig. 3.8). Hence the effect on S should

3.6 Controlling Nucleation Rate, Number Density and Size of PBLG Liquid Crystals

dominate over that on σ.

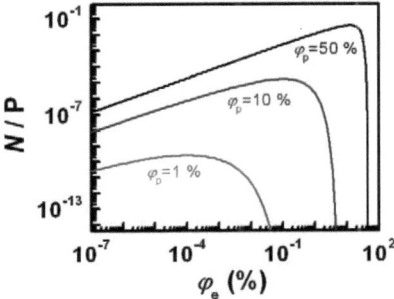

Figure 3.8: Relative number density N/P of nuclei as a function of equilibrium volume fraction φ_e at three different polymer volume fractions $\varphi_p > \varphi_e$ according to 2.20 with $\varphi'_e = 1$.

Fig. 3.8 shows that the dependence of the interfacial tension σ and the supersaturation ratio S on the equilibrium volume fraction φ_e affects nucleation rate and number density in two opposite ways. The result of these effects is a maximum in nucleation rate and number density at a specific equilibrium volume fraction $\varphi_e = \varphi_P^3/\varphi_e'^2$.

Fig. 3.9 depicts the evolution of the object area (a) and length (b) with time as

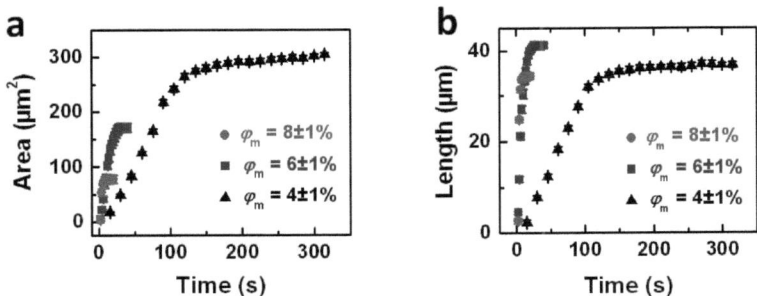

Figure 3.9: Evolution of area (a) and longest length (b) of liquid crystalline objects in time measured for objects marked with corresponding colored ellipses in Fig. 3.6. The length was measured along the long axis of the objects.

observed by OM during the growth of the marked objects in Fig. 3.6 with corresponding colored ellipses. The objects area is larger for those, which were grown in PBLG film solutions containing lower amounts of methanol, i.e. for lower Nh and more available PBLG per nucleus.

Figure 3.10: An optical micrograph showing a grown PBLG liquid crystal. The experiment was like that of Fig. 3.6. Continuous adding methanol into the thin film solution led to a continuous but slow decrease of φ_e which allowed supersaturation ratio S to be bigger than 1. If $S > 1$ for a long time the object continuously grows to a large size with the length of about 207 μm. Size of the micrograph is $330 \times 264\,\mu\mathrm{m}^2$.

Nonetheless, the final length of the objects was found to be nearly equal and independent of φ_m. During the growth, the PBLG volume fraction in the isotropic part of the mixture decreased, causing a decrease in growth rate. Eventually, when the solubility limit was reached ($\varphi_\mathrm{p} = \varphi_\mathrm{e}$), the objects stopped growing. In additional experiments (see Fig. 3.10), we used the advantage of our new technical approach in order to extend the growth process through a continuous but slow increase of the amount of condensed methanol during growth via a continuous and gentle increase of the methanol flow rate. i.e. a continuous but slow decrease of φ_e. this allowed us to obtain hexagonal columnar liquid single crystals as long as about 200 μm.

Fig. 3.11a shows the evolution of N with polymer volume fraction φ_p for three different methanol volume fractions φ_m present in the thin film solution. Here data from previous studies [28, 61, 96] on solutions of PBLG star block copolymers in chloroform with trace amounts of methanol were included. The reason for including these data is that similar behavior of nucleation, growth, structure and pattern formation as well as the same effect on φ_e by adding protic nonsolvents in their solution were observed. Also, fittings to these data were in good agreement with

3.6 Controlling Nucleation Rate, Number Density and Size of PBLG Liquid Crystals

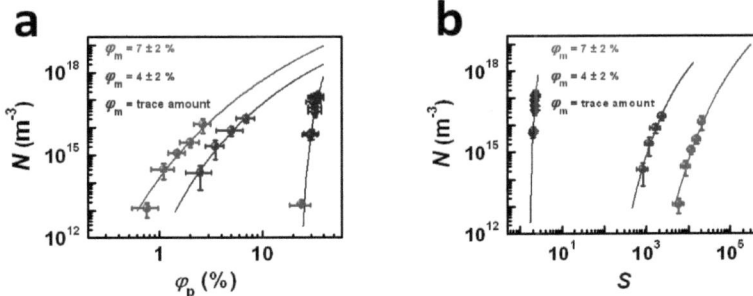

Figure 3.11: Number density of nuclei as a function of (a) polymer volume fraction φ_p and (b) supersaturation ratio S obtained when adding different amounts of methanol to the thin film solution. The solid lines are corresponding fits to equation 2.20 by assuming $Q = \frac{3}{5}$ for a diffusion controlled growth process, $\beta = \frac{16\pi}{3}$ for spherical nuclei (β is a shape factor introduced by Nielsen to account for different nucleus geometries [52]), $v = \frac{4}{3}\pi r^3$ (which is the volume of a spherical molecule of radius r) and $\sigma = \frac{kT}{2\pi r^2} \ln\left(\frac{1}{\varphi_e}\right)$ [52]. The obtained values of φ_e and P are represented in Tab. 3.1. The dark blue symbols in a and b are obtained from previous studies [28, 61, 96] and represent data for chloroform solutions of rod-coil star block copolymers with PBLG forming the rod blocks, containing trace amounts of methanol. The dark circles in represent results taken from previous investigations [28, 61, 96] on chloroform solutions of rod-coil block PBLG copolymers, containing only trace amounts of methanol. The green star in (a) represents the result of an analogous experiment on a chloroform solution of the here studied PBLG homopolymer, containing only a trace amount of methanol. As can be seen in (a), a fit of 2.20 to this data point in combination with results for PBLG copolymers yields satisfactory agreement.

our results obtained from PBLG homopolymers in that range of polymer volume fraction and methanol content. For these reasons, the results of previous works are good candidate to compare with the results obtained from PBLG homopolymers.

Values of the equilibrium volume fraction φ_e (see Tab. 3.1) were obtained by non-linear parameter fits to equation 2.20, and were found to be roughly $1.3 \times 10^{-4}\%$ (for $\varphi_m = 7 \pm 2\%$), $3.0 \times 10^{-3}\%$ (for $\varphi_m = 4 \pm 2\%$) and 14.6% (for trace amounts of methanol).

Using the values of φ_e from Tab. 3.1 allowed us to plot the number density N as a function of the supersaturation ratio S (Fig. 3.11b) for the three different amounts

φ_m (%)	φ_e (%)	P (m^{-3})
7 ± 2	$(1.3 \pm 1.8) \times 10^{-4}$	$(6.1 \pm 1.8) \times 10^{23}$
4 ± 2	$(3.0 \pm 1.8) \times 10^{-4}$	$(15 \pm 2.9) \times 10^{21}$
Trace amount	14.6 ± 1	$(9.9 \pm 5.3) \times 10^{19}$

Table 3.1: This table shows the parameters obtained by nonlinear fits according to the solid lines in Fig. 3.11a and Fig. 3.11b

of methanol present in the film solution in agreement with the theoretical plots according to 2.20 (solid lines in Fig. 3.11b).

In Fig. 3.12 the dimensionless quantity $\dfrac{(Q\beta v^2)^{1/3}\sigma}{kT}$, representing the interfacial tension σ between the nuclei and the surrounding solution, is shown as a function of the equilibrium volume fraction φ_e. The solid red line represents the dependence on the base of the theory of Nielsen [52] and theory of Mersmann [56], as used in

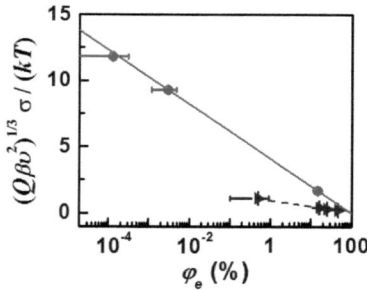

Figure 3.12: Interfacial tension σ between nuclei and surrounding liquid phase, represented by the dimensionless quantity $\dfrac{(Q\beta v^2)^{1/3}\sigma}{kT}$ (see 2.19 and [52]), as a function of the equilibrium volume fraction φ_e consistent with previous studies [52]. the solid line corresponds to the dependence used in 2.20. The dark blue triangles represent results for previous investigations [28,61,96] and represent data for chloroform solutions of rod-coil star block copolymers with PBLG forming the rod blocks, containing different amounts of water as the protic non-solvent for PBLG.

equation 2.20. The dark blue symbols represent, for comparison, the corresponding interfacial tension obtained in previous studies [28,61,96] on chloroform solutions of

PBLG star block copolymers with various amounts of water added, using a similar way of analyzing the data. The best fit yields a slope of -0.52 ± 0.01. Differences in the values of the slopes for the here studied homopolymer and the previously studied star block copolymer are attributed mainly to the differences in the nonsolvents used and also to uncertainties in determining the values of Q, β, v and a_m.

The observation that the nonsolvent influences the solubility limit or equilibrium volume fraction was explained in two different ways: calculations [23] on ternary systems of PBLG, solvent and nonsolvent on the base of Flory's theory [18] show that the equilibrium volume fraction of PBLG in the isotropic phase can strongly decrease with increasing volume fraction of the nonsolvent. This is in consistence with experimentally obtained phase diagrams [48,99]. Alternatively, we may assume the formation of a complex between polypeptide molecules and protic nonsolvent molecules like water or methanol [96,100]. These complexes have a different solubility. We tried to check for such possible complexation between methanol and PBLG molecules by NMR measurements (see sec. 3.7).

3.7 NMR Investigation on PBLG Methanol Complexation

In order to check for possible complexation between a protic nonsolvent (methanol in this case) and PBLG molecules we did NMR measurements on both a reference sample containing PBLG molecules dissolved in dimethylformamide (DMF) (10 mg/0.8 ml) and a sample containing PBLG, DMF and methanol (10 mg/0.8 ml/0.2 ml) using 1D NMR as well as 2D proton–proton correlation (Nuclear Overhauser Effect (NOE)) methods. In this series of experiments, we used DMF instead of chloroform because in contrast to chloroform the loss of DMF from the standard NMR samples that were used will be negligible during the time of the NMR experiments.

All spectra were acquired at a Bruker Avance II$^+$400 MHz spectrometer using a 5 mm BBFO probehead. The decoupler coil was tuned to the proton frequency of 400.17 MHz. The lengths of the 90° impuls was 14.1 us. The chemical shifts are given with respect to TMS.

The ^{1}H – NOESY spectrum was acquired at a spectral width of 4 kHz. 2048 data points were collected in the t_2 domain and 256 time increments in the t_1 domain. 48 scans were sampled for each FID. Data acquisition and processing were carried out in the TPPI mode. Squared sinusoidal functions were applied in both dimensions.

The size of the final matrix was 1024 × 1024.

Fig. 3.13a shows a schematic representation of the PBLG molecule. All parts of the molecule that are represented by resonances in the ^{1}H − NMR spectra (Fig. 3.13b) are marked with numbers from 1 to 6. Unmarked significant signals come from hydrogen atoms of the solvent (DMF). Resonances of some impurities or water molecules present in the solution can be distinguished as well but they are very weak in comparison to the signals of solute and solvent molecules.

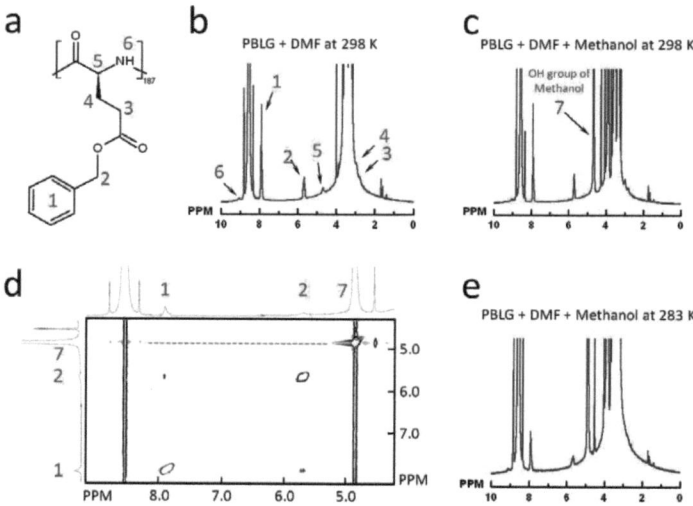

Figure 3.13: (a) Schematic representation of PBLG molecule. All parts of the molecule that are represented by resonances in the ^{1}H − NMR spectra are marked with numbers 1 to 6. (b) NMR spectra of the PBLG / DMF solution. Each peak coming from PBLG is marked with corresponding number in a. (c) NMR spectra of the isotropic PBLG / DMF / methanol solution at 298 K. (d) NOE or proton–proton correlation measurement of a PBLG / DMF / methanol solution at $T=$ 298 K. (e) NMR spectra of the PBLG / DMF / methanol mixture, phase separated at $T=$ 283 K.

Fig. 3.13c represents the NMR spectra of the PBLG / DMF / methanol isotropic solution at $T=$ 298 K. Comparing the spectra with Fig. 3.13b and spectra of the same sample at $T=$ 313 K (not shown here), the resonances coming from PBLG

3.7 NMR Investigation on PBLG Methanol Complexation

remained unchanged with the exception that the signal (5) coming from PBLG is hidden by the strong resonance corresponding to the OH group of methanol.

Fig. 3.13d shows the ^{1}H − NOESY spectrum of a PBLG / DMF / methanol solution. Crosspeaks in this spectrum indicate the presence of a NOE between the different spins whose signals are connected by the correlation. The NOE results from magnetization transfer by crossrelaxation between chemically inequivalent spins. Crossrelaxation processes depend on the dipole-dipole relaxation mechanism that is based on the dipole-dipole interaction through space between neighboring magnetic moments. Because the magnitude of this dipole-dipole interaction is highly related to the distance of the interacting spins, the presence of a NOE indicates close spatial neighborhood of these spins. NOEs can occur not only intramolecular but also intermolecular if spatial neighborhood of different molecules is statistically frequent and stays long enough for magnetization transfer to built up. In the ^{1}H − NOESY spectrum of the PBLG solution all intramolecular NOEs that can be expected in PBLG are observed, e.g. between the protons of the phenyl group (1) and the protons of the neighboring CH_2 group (2). In contrast no intermolecular NOE is found between signals of PBLG and the resonances of methanol. This argues against a close interaction of PBLG and CH_3OH in this special solution. Nonetheless, we do not exclude the possibility for such a bonding, noting that intermolecular NOEs can get quite small and may be not detectable if the number of interacting molecules is as low as in this example.

By decreasing the temperature of the solution 283 K, phase separation into a solution with low volume fraction φ_p of PBLG and an anisotropic phase with high φ_p of PBLG took place, and therefore the NMR resonances corresponding to PBLG drastically decreased (Fig. 3.13e), as the spectrum shows only contributions of the PBLG in the isotropic phase of the solution, because the anisotropic phase is not in the active volume of the NMR measurement. By integration of the ^{1}H − NMR spectrum one can investigate if the concentration of methanol is also reduced.

NMR is an intrinsically quantitative method because the intensity of a NMR signal is directly proportional to the number of spins that induce the signal. Of course, the integrals I are relative and are compared with a reference signal. To investigate if the concentration of methanol is reduced this relative intensities as well as the determination of absolute concentrations are very helpful. In the spectrum at hand the signal of the CHO group of the solvent DMF at 8.59 ppm is chosen as reference signal and its integral is deliberately set to the absolute value 1. The signals of the phenyl group at 7.93 ppm, 7.92 ppm and 7.87 ppm are chosen to represent the PBLG because they are the most intensive ones of this molecule and not much affected by overlapping parts of other signals. Finally for methanol the resonance of the OH group at 4.66 ppm was integrated. This resonance overlaps with the signal of the CH group 5 of PBLG. For the correct integral of methanol the contribution of the latter is subtracted. To get the molar ratios of the molecules out of the integrals I, it must be taken into account that the phenyl group contains 5 protons whereas the groups chosen for the solvent and nonsolvent contain only one proton. To get the relative number of molecules n_i the integrals I_i must be divided by the number of protons N_i that are contained in the group whose signal is integrated:

$$n_i = \frac{I_i}{N_i} \tag{3.1}$$

For PBLG the number of phenyl groups doesn't present the number of polymer molecules but presents the number of monomer units in solution. Three ^{1}H − NMR spectra were measured at different temperatures delivering the data represented in Tab. 3.2.

T (K)	DMF		Methanol		PBLG monomer unit	
	I_{DMF}	n_{DMF}	$I_{methanol}$	$n_{methanol}$	I_{PBLG}	n_{PBLG}
313	1	1	0.442	0.442	0.022	0.0044
298	1	1	0.441	0.441	0.022	0.0044
283	1	1	0.440	0.440	0.007	0.0014

Table 3.2: The values obtained from integration over spectra from three components of the mixture at three different temperatures. The integration is done over CHO group of DMF, phenyl group of PBLG and OH group of methanol corrected for the contribution of CH group 5 of PBLG.

As to be expected the chemical shift of the proton signals changes slightly with temperature but the selected resonances stay well separated from each other as well as from other signals in the spectrum. Therefore the quality of the integration is not affected.

At 313 K and 298 K the ratio of the integrals of the three molecules is the same (see

3.7 NMR Investigation on PBLG Methanol Complexation

Tab. 3.2). In solution for each molecule DMF there are 0.45 molecules of methanol and 0.0044 monomer units of PBLG. Even with a hundredfold excess of methanol molecules per monomer unit the PBLG polymer stays in solution at 298 K. When phase separation happens and anisotropic phase is formed at 283 K the amount of PBLG in solution is reduced by about 70 %. Because of the huge excess of the solvent molecules DMF and methanol, deviations of the composition of the solvents incorporated in the precipitate will not affect the integral ratio in solution significantly. Under the given circumstances an error of about 10 % must be assumed for the integrals and the molar ratio. This means that if for example the precipitate does not contain any DMF, there must be at least 13 methanol molecules per monomer unit of the PBLG in the solid to bring about a significant decrease of the methanol integral in solution. At least this constitutes an upper limit for the methanol content of the precipitate.

By using the results of the integrals one can obtain the values of the molar fractions x_i in the isotropic solution:

$$x_i = \frac{n_i}{\sum n_i} = \frac{I_i / N_i}{\sum I_i / N_i} \tag{3.2}$$

Integrating over the NMR signals at 313 K, 298 K (before phase separation) and 283 K (after phase separation) gave us the equilibrium values for $x_{i,e}$ in the isotropic part of the phase separated mixture represented in Tab. 3.3.

T (K)	$\sum n_i$	x_s	x_m	x_p
313	1.4464	0.69±0.02	0.31±0.02	0.0030±0.0003
298	1.4464	0.69±0.02	0.31±0.02	0.0030±0.0003
283	1.4464	0.69±0.02	0.31±0.02	0.0010±0.0001

Table 3.3: Molar fractions of DMF, methanol and PBLG in the isotropic mixture at three different temperatures.

Here indices p, s and m denote polymer (PBLG), solvent (DMF) and nonsolvent (methanol) respectively. x_p here is the molar fraction of the PBLG monomer units, not of the entire polymer.

Using NMR, we could get information only on the isotropic phase. Therefore, we have no direct and clear information about possible bonding or correlation between PBLG and methanol molecules in the anisotropic phase. NMR results confirm our PBLG thin film solution studies that addition of methanol cause phase separation in the mixture. No direct interaction between protons of methanol and PBLG molecules was detected in the isotropic solutions. However, the fact that at 298 K there is an excess of methanol molecules to PBLG monomers by hundredfold and still no phase separation has taken place can be a hint that strong hydrogen bonding between methanol OH group and PBLG side chains is less plausible. This is in consistence with infrared spectroscopic results [100], where dichroic OH stretching absorption of water and alcohol molecules were found for some sufficiently hydrophilic polypeptides, but not for the more hydrophobic poly(γ-benzyl glutamate).

3.8 Conclusion

In summary, we have demonstrated a general method for extremely fine control of nucleation, growth and dissolution of PBLG hexagonal columnar liquid crystals in thin film solutions of low polymer concentration by adding and removing different amounts of methanol. This method can also be applied for other type of molecules like for example semiconducting polymers or bio polymers. By this method we are able to control the rate of nucleation and the number density of nucleated objects. We showed that adding the nonsolvent methanol to the isotropic polymer solutions decreases the equilibrium volume fraction and hence, promoted nucleation and growth of PBLG hexagonal columnar liquid crystals. The laws derived from classical nucleation theory for crystals were found to fit well for nucleation and growth of the liquid crystalline structures. NMR experiments gave no evidence for a complexation between PBLG and methanol molecules in the isotropic phase while it is not a proof against such complexations. In fact, an interaction of the PBLG and methanol molecules in the anisotropic phase could not be ruled out on the basis of the performed measurements. The drastic decrease of the equilibrium volume fraction with increasing content of methanol, however, can be well understood on the base of the Flory theory as the effect of a nonsolvent component.

4 Structure, Pattern Formation and Orientation of PBLG Molecules within the Crystals

4.1 Introductory Remarks and State of the Art

Similar to natural polypeptides, synthetic polypeptides like poly(γ-benzyl L-glutamate) (PBLG) can adopt an α-helical conformation [29] in the solid state [30] as well as in helicogenic solvents [31, 32]. The α-helical polymer backbone, built up by the amide groups, is stabilized by intra-molecular hydrogen bonds [29]. When formed by the residues of L amino acids (as polypeptides in nature), the helix is right-handed [33]. Polypeptide molecules of α-helical conformation can be regarded as stiff, rod-like particles. For solutions of such particles with a sufficiently high aspect ratio, liquid crystalline states have been predicted theoretically, when the volume fraction of the rods exceeds certain limits [18, 90]. These theories considered only transitions to the nematic phase, which is the simplest liquid crystalline state. They were partially confirmed by experimental data [34, 39]. However, the systems can show a very complex behavior, including the occurrence of different liquid crystalline states, whereas the theories of Onsager and Flory do not treat details of the states beyond the existence of orientational order. In particular, a chiral nematic phase [34] and, at high volume fractions of the polypeptides, a hexagonal columnar liquid crystalline phase [22] have been found. The latter phase possesses a solid-like hexagonal order in two dimensions and a liquid-like in the third [94]. Corresponding hexagonal lateral order of the PBLG rods, with a spacing correlated to the volume fraction of solvent in the anisotropic phase, was already assumed in earlier studies [23, 101]. Within this complexity of the phase diagram, we focused only on the highly ordered phases (see chapter 3) for which we have adopted an experimental approach allowing to microscopically follow the crystallization process in real time and in direct space (see sec. 3.2.2). As a result, we obtained large liquid crystals of PBLG in solution. In the solid state (i.e., without any solvent), rod-like polypeptides can crystallize in different modifications. These structures depended on the solvent used and the way the solid state has been prepared. Monoclinic, hexagonal and pseudohexagonal crystal structures have been identified in bulk samples and films of PBLG [21, 37, 92, 102]. In this thesis, we obtained PBLG liquid crystals by nucleation and growth in semi-dilute thin film solutions. They turned to dry PBLG crystals after drying and have been subsequently characterized by different means.

Crystallizing PBLG molecules from thin film solutions allowed to employ optical microscopy for studying nucleation and growth of single, randomly oriented objects in real time. In previous chapters as well as in studies of rod-coil block copolymers containing PBLG blocks as the rod parts [61, 96] and also of a PBLG homopolymer [96], it was shown that the resulting objects were of anisotropic shape containing domains comprised of parallel stripes that were changing their orientation alternately, resulting in a zig-zag pattern. So far, the internal order of these objects was not determined, i.e. it was not clear if these were crystalline objects. In order to allow investigations by electron diffraction and X-ray scattering, which provide information on order and orientation of PBLG helices within these objects, it was necessary to obtain objects large enough for such studies. In the previous chapter we explained the method of getting such crystals.

In this chapter we want to determine the lattice structure of PBLG objects using electron diffraction and X-ray scattering methods and determine the orientation of PBLG molecules with respect to the plane of the substrate and within the crystalline structure. The birefringency of the crystals will be exploited and finally the origin of the zig-zag patterns observed on the surface of the PBLG crystals will be discussed.

4.2 Transformation from Single Domain to Multi Domain

Both, the experimental setup and technique used to create a thin film solution, in which nucleation and growth of PBLG crystals could be studied in situ, were described in the previous chapters (see sec. 3.2, sec. 3.3 and sec. 3.6). After growing the PBLG crystals to a sufficiently large size, thin film solutions were dried by rapidly evaporating both solvent and non-solvent molecules, simply by increasing rapidly the sample temperature up to 65 °C, where it was kept for a few minutes

4.2 Transformation from Single Domain to Multi Domain

(see Fig. 4.1).

Fig. 4.1a depicts an optical micrograph of a typical PBLG liquid crystal in a

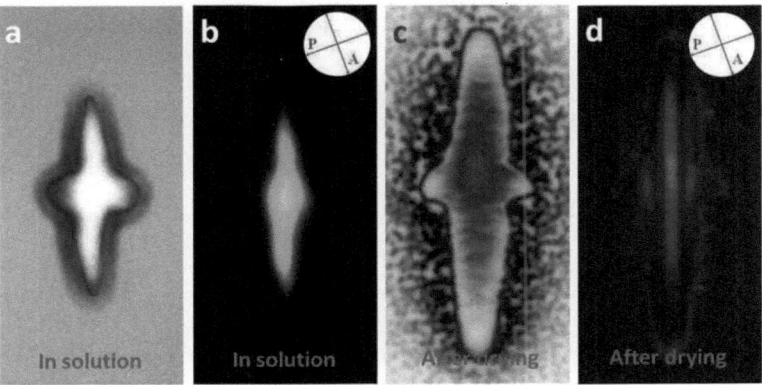

Figure 4.1: Optical micrographs of a grown PBLG crystal still in solution, taken in reflection mode (a) without and (b) with crossed polarizers. Between crossed polarizers just one birefringent bright domain could be detected. (c) Optical micrograph of a dried PBLG crystal in bright field reflection mode (the crystal was grown under the same conditions as the ones in Fig. 3.4 and Fig. 4.1a. Different interference colors indicate variations in thickness of the crystal. (d) When using crossed polarizers, the dried PBLG crystal exhibits several birefringent domains of alternating brightness, aligned parallel to each other. The size of all images is $14 \times 30 \,\mu m^2$.

film solution, i.e. before drying. Even when surrounded by solution, this PBLG liquid crystal showed birefringence over about the same size and shape, as revealed under crossed polarizers (Fig. 4.1b), suggesting a single domain crystal. In contrast, after drying such PBLG crystals (Fig. 4.1c), this single domain always broke up into several domains; separated by straight boundaries and each exhibiting uniform birefringence. However, these approximately parallel domains showed alternating brightness (Fig. 4.1d). We emphasize that the breaking up into multiple domains was only observed after the drying process and does not depend on the morphology of the liquid crystals i.e. both single domain cross like and denderitic crystals portioned into multiple birefringence domains (see Fig. 4.2). While embedded in the solution film, all PBLG crystals showed only one single domain.

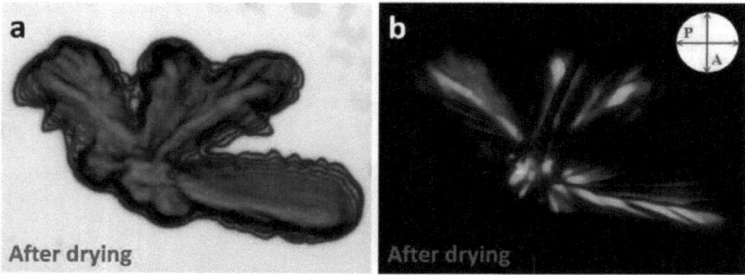

Figure 4.2: (a) Optical micrograph of a dried PBLG crystal with denderitic morphology in bright field reflection mode (the crystal was grown under the same conditions as the one in Fig. 3.4. (b) When using crossed polarizers, the dried PBLG denderitic crystal exhibits several birefringent domains just like crystals with other type of morphologies. The size of all images is $185 \times 140\,\mu m^2$

Dried PBLG crystals were then characterized using optical microscopy with polarized light (POM), atomic force microscopy (AFM), X-ray and electron diffraction methods.

In order to identify the structure of the PBLG lattice and the orientation of the molecules within the crystals, wide angle X-ray scattering (WAXS) measurements were performed. We used a modified Siemens D500 X-ray diffractometer with a conventional Cu-Kα X-ray source of wavelength $\lambda = 0.1542$ nm. A thin film containing many large but randomly orientated PBLG crystals, all lying flat on the silicon substrate, was chosen as a representative sample.

Electron diffraction measurements were performed on a Zeiss (LEO) 912 Omega electron microscope and diffractometer operated at an acceleration voltage of 120 kV, using the software SIS Olympus iTEM. PBLG crystals were transferred from the silicon substrate onto an appropriate TEM grid, keeping the same orientation of PBLG crystals as on the initial silicon substrate. For that, we have grown PBLG crystals from thin films spin-cast onto silicon substrates covered with a water soluble layer of poly(3,4-ethylenedioxythiophene):poly(styrenesulfonate) (PEDOT:PSS). After growing and drying PBLG crystals on such substrates, these solid samples were exposed to water. As PEDOT:PSS dissolves in water, the grown PBLG crystals were lifted off the silicon substrate. The crystals floating on the water surface were then picked up with electron microscopy copper grids coated with carbon.

4.3 Surface Topography Measured by AFM

Figures Fig. 4.3a and Fig. 4.3c depict AFM height and phase images of the same dried PBLG crystal shown in Figures Fig. 4.1c and Fig. 4.1d. The height variations within the PBLG crystal range from 150 nm at the edges to about 300 nm at the center. We note that the thickness of the crystal is small compared to its lateral dimensions.

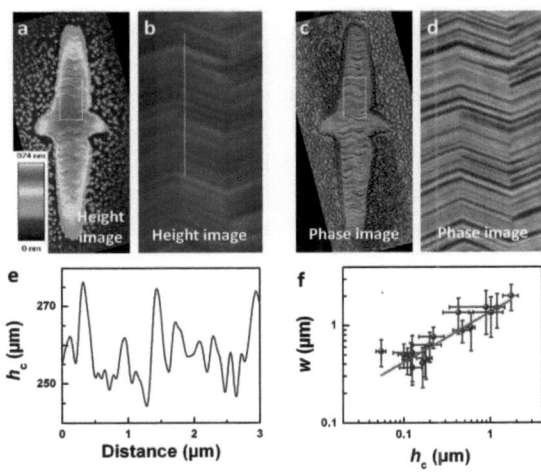

Figure 4.3: (a) AFM height and (c) phase images of the dried PBLG crystal shown in Figure Fig. 4.1c. The black triangular regions originate from rotations of the original AFM image to align the crystal axis vertical and cutting the edges. (b, d) Domains of parallel stripes, alternately changing their orientation leading to zig-zag patterns, can be clearly observed on top of PBLG crystals. The area shown in (b, d) is indicated by a white rectangle in (a, c). These domains of parallel stripes seen by AFM coincide with the birefringent domains in Figure Fig. 4.1d. (e) A typical height profile along the vertical white line in (b), h_c denotes the thickness of the crystal. (f) Dependence of the width of the birefringent domains w on the maximum thickness of PBLG crystals h_c. The straight line corresponds to a fit to a power law $w \sim (h_c)^\beta$ with an exponent $\beta = 0.52 \pm 0.06$. The size of the images in (a) and (c) is $14 \times 30\,\mu m^2$ and in (b) and (d) it is $2.5 \times 5\,\mu m^2$.

As can be clearly seen in the phase contrast image (see Fig. 4.3c), but also visible in the height image (see Fig. 4.3b), each of the domains visible in the birefringence image of Fig. 4.1d consisted of uniquely oriented parallel stripes, which can be clearly distinguished on the surface of the crystal. Between adjacent domains all stripes changed conjointly their orientation, generating a zig-zag pattern. Three such domains are shown in Fig. 4.3b (height image) and Fig. 4.3d (phase image). The change in direction is rather similar between all domains, yielding an average angle of these zig-zag kinks of 42° ± 4° in this PBLG crystal. However, for different PBLG crystals, this angle can be different; but the deviation from the individual average for each crystal is still 4° to 5°. Averaging over many PBLG crystals showed that the range of these kink angles is 35° ± 15°. Fig. 4.3e represents a height profile along the vertical white line shown in Fig. 4.3b, indicating the steps in height between parallel stripes. At this point, we cannot exclude that these steps were generated by the drying process of PBLG crystals. Fig. 4.3f shows that the width w of the domains (in the direction normal to the separating boundaries) increased with the mean thickness h_c of the PBLG crystals. This increase can be described well by a power law w $\sim (h_c)^\beta$ with an exponent $\beta = 0.52 \pm 0.06$.

4.4 Diffraction Measurements

4.4.1 X-Ray Measurement

During the WAXS measurement the angle of the incident beam with respect to the substrate plane was θ. Both, the sample and the detector were moved simultaneously in a way that the angle between incident and reflected beam was always equal to 2θ (see Fig. 4.4a). As a result, Fig. 4.4b shows the WAXS intensity as a function of scattering angle obtained from a sample containing many PBLG crystals (thickness $h_c > 1\,\mu$m), all lying flat on the silicon substrate but being randomly oriented within the plane of the substrate. The WAXS curve exhibits a peak at $2\theta = 7.18° \pm 0.03°$, corresponding to a periodicity $d = \lambda/(2\sin\theta) = 1.23 \pm 0.01$ nm. To obtain information about the orientational distribution of the lattice planes, we performed additional scattering experiments. We scanned the angle 2θ between scattered and incident beam at several fixed angles ω between incident beam and surface plane in the range from 1° to 6°. We observed pronounced scattering at $2\theta \approx 7.2°$ only at angles of incidence very close to half of the diffraction angle 7.2°, as shown in the Fig. 4.4c (for reasons of clarity curves are shown only for three different angles ω from 3° to 4°). Accordingly, we assume that a pronounced portion of the corresponding lattice planes are oriented parallel to the substrate plane. In previous experiments

4.4 Diffraction Measurements

from other groups [21, 102] a slightly deformed hexagonal lattice was found for PBLG crystallized from chloroform solutions. Our experimentally obtained value for d is close to the reported value of 1.26 nm to 1.28 nm for the (010) plane of the near-hexagonal lattice of the α-helical PBLG crystals, deduced from X-ray scattering experiments (see Fig. 4.4d) [21].

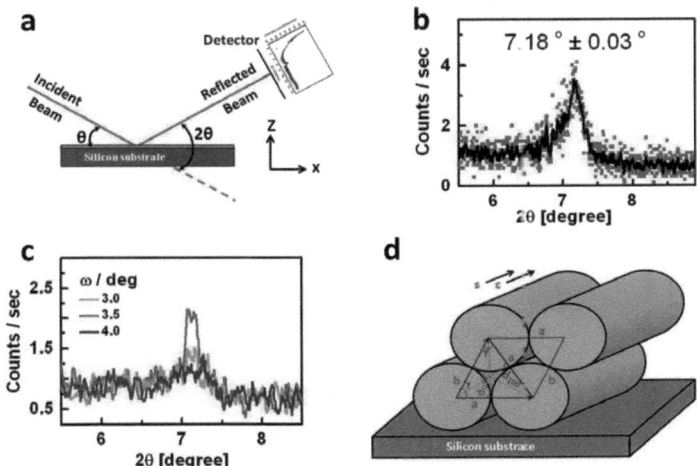

Figure 4.4: (a) A schematic representation of the set up used for WAXS measurement in reflection mode. Incident angle θ is the angle between incident X-ray beam and substrate surface. 2θ is the angle between incident and reflected beam and is two times of incident angle. (b) WAXS intensity obtained in reflection mode ($\theta / 2\theta$ scan) emphasizing a peak at $2\theta = 7.18° \pm 0.03°$ that is related to the periodicity of $d_{010} = 1.23 \pm 0.01$ nm. (c) WAXS intensity obtained in reflection mode for several fixed angles of incidence ω. (d) Schematic of the deduced and possible arrangements of α-helices in a pseudohexagonal lattice structure with respect to the substrate from X-ray measurement. The vector **s** denotes the direction of the stripes seen in the AFM experiments.

4.4.2 Electron Diffraction Measurement

In order to confirm the near-hexagonal packing and to identify lattice constants, we have performed complementary electron diffraction measurements in transmission mode (see Fig. 4.5a) which revealed a periodicity of 1.33 ± 0.03 nm. This distance is also compatible with a near-hexagonal packing of PBLG helices within the crystals. We note that in contrast to an incident angle of the X-ray beam almost parallel to the substrate, the electron beam was passing through a crystal with multiple domains at an angle orthogonal to the plane of the electron microscopy grid. The obtained periodicity is comparable to the value of 1.28 nm to 1.34 nm reported for the (100) plane of the pseudo-hexagonal lattice of the PBLG crystals [21]. Based on the measured values of electron diffraction and X-ray we thus took $d_{100} = 1.33 \pm 0.03$ nm and $d_{010} = 1.23 \pm 0.01$ nm respectively. Following trigonometric considerations (see Fig. 4.4d and Fig. 4.5b), this results in an angle $\gamma = \arccos(d_{010}/2d_{100}) = 62.5° \pm 0.7°$ between the lattice vectors and in lattice constants $a = d_{100}/\sin\gamma = 1.50 \pm 0.03$ nm and $b = d_{010}/\sin\gamma = 1.39 \pm 0.02$ nm.

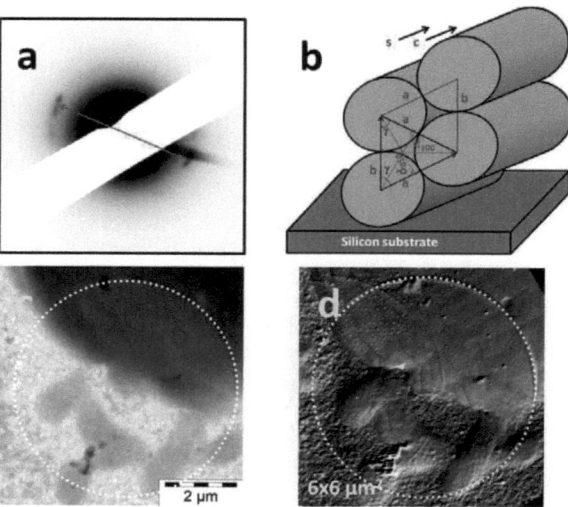

Figure 4.5: (a) Electron diffraction pattern obtained from a PBLG crystal. The red arrow indicates the periodicity of $d_{100} = 1.33 \pm 0.03$ nm. (b) Schematic of the deduced and possible arrangements of α-helices in a pseudohexagonal lattice structure with respect to the substrate from TEM measurement. The vector s denotes the direction of the stripes seen in the AFM experiments. Bright field TEM (c) and AFM phase image (d) showing the circular area, where the electron beam affected the sample during the diffraction measurement of (a).

In literature, the surface of PBLG crystals obtained from solutions in chloroform

4.4 Diffraction Measurements

was reported to be parallel to the ($\bar{1}$20) lattice plane, according to the α-helical rods lying in the surface plane [21]. Also in a previous report on a related system of a rod-coil block copolymer containing PBLG rod blocks [103], they were assumed to be oriented parallel to the substrate plane. The electron diffraction spots originating from multiple domains (Fig. 4.5d) within a single dried crystal are located on a small arc of a circle (see Fig. 4.5a). Due to the lability of the crystal to the electron beam, diffraction measurements were performed in the so-called MDF-Mode (Micro Dose Focusing), i.e. the electron beam is focused outside the measurement area and the diffraction pattern is taken by jumping to the measurement area without further focusing. After this, the observed diffraction pattern vanishes within seconds, indicating damage of the crystal structure due to the electron beam. Hence, bright field TEM image (Fig. 4.5c) were taken after measuring the diffraction pattern. In the bright field TEM (Fig. 4.5c) and AFM phase image (Fig. 4.5d) (the area of the electron beam used at the diffraction experiment is indicated by the white dotted circle), a strong perturbation of the crystal is seen due to interaction with the electron beam. However, a trace of domains with different orientation of the stripes still can be distinguished in the AFM phase image (Fig. 4.5d). We assume that the opening angle of the arc is due to an orientational distribution of the lattice vector **c** for the different domains within the surface plane. The orientation of the lattice with the ($\bar{1}$20) lattice plane parallel to the surfaces of the crystal is depicted in the Fig. 4.5b. The observation of the reflections at the (100) planes by the electron diffraction experiment corresponds well to this orientation: To meet the Bragg condition for the very small scattering angle, corresponding to the very short de Broglie wavelength of the electrons (3.35 pm at 120 kV), the vertically oriented electron beam should be nearly parallel to the (100) planes. This is corroborated by the absence of the d_{100} reflections in the X-ray diffraction experiment, which would show up at $2\theta = 6.6°$ in Fig. 4.4a. Obviously, the Bragg condition for the reflection at the (100) planes is nearly never met in the X-ray experiment, which mainly detects reflections at lattice planes parallel to the substrate. However, as indicated in the Fig. 4.5b, the (010) planes are tilted by $\gamma/2$ with respect to the substrate, hence not meeting the Bragg condition for the X-Ray diffraction. We assume that at least a sufficiently large portion of subdomains is oriented as shown on the left side of Fig. 4.4d, with the (010) planes parallel to the substrate and hence meeting the Bragg condition for the X-ray diffraction experiment. The AFM phase contrast of neighboring stripes shown by Fig. 4.3e might indicate a corresponding distribution of the rotation of the (100) and (010) planes around the **c** axis.

4.5 Birefringent Domains and Molecular Orientation

After finding the orientation of **c** axis of pseudohexagonal lattice with respect to the substrate, now we use optical properties of multi domain crystals combined with results of AFM, in order to find the orientation of **c** axis (and hence, long axis of PBLG molecules) with respect to the stripes within the birefringent domains. We used POM in transmission mode for these series of experiments (see Fig. 4.6). A λ plate compensator (retarder) used in this setup as well in order to distinguish between fast and slow optical axis of birefringent domain.

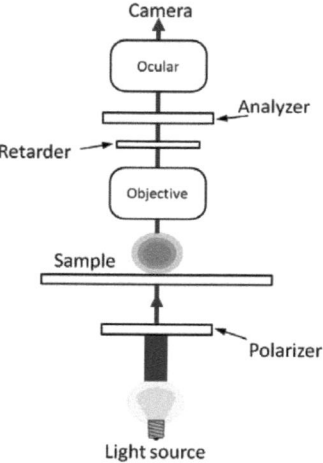

Figure 4.6: A schematic representation of the POM set up used in order to find the orientation of the **c** axis of pseudohexagonal lattice with respect to the stripes within the birefringent domains.

Fig. 4.7 shows birefringent domains of a PBLG crystal observed by transmission mode POM. In order to transfer PBLG crystalline objects onto a transparent glass substrate, we used the same technique as used to transfer crystals onto TEM grids.

Using crossed polarizers and consistent with Fig. 4.3, ten to eleven birefringent crystalline domains showed up (Fig. 4.7a to Fig. 4.7d), consisting of parallel stripes as

4.5 Birefringent Domains and Molecular Orientation

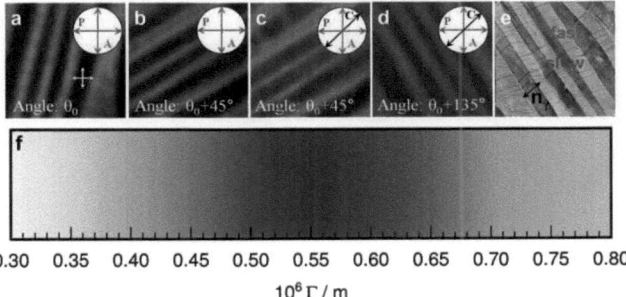

Figure 4.7: (a), (b), (c) and (d) POM images of a PBLG crystal, observed in transmission mode, using crossed polarizers. The insets show the orientations of polarizer P, analyzer A and λ plate C. In (a), five of the domains appear dark, indicating that their optical axes are parallel to polarizer or analyzer. In (b) the crystal was rotated by 45° with respect to (a); domains, which were dark in (a) appear now bright. In (c) a λ plate, oriented with its slow axis C diagonal to the polarizers (see inset), was used. As a result, the bright domains of (b) show up in an orange color, indicating that their slow axes were about perpendicular to that of the λ plate, resulting in a subtraction of the retardations. In (d) the crystal was rotated 90° more with respect to (c); as a result the bright domains of (b) show up in a blue color, indicating that their slow axes were about parallel to that of the λ plate, resulting in an adding-up of the retardations. In (e) an AFM phase image corresponding to (d) is shown, indicating that the slow optical axes of the domains s, were parallel to the stripes. (f) Interference color gradient for a birefringent object placed in diagonal orientation between crossed polarizers calculated for a light source corresponding to a black body radiator at a temperature of 6500 K (color temperature of daylight) and optical path differences Γ around that of a λ plate (0.55 μm). The size of the images from a to e is $8 \times 8 \, \mu m^2$.

revealed by the corresponding AFM phase image (Fig. 4.7e). By rotating the crystal with respect to the polarizers, the domains could be brought into orientations where they appeared dark (normal orientation) (see Fig. 4.8). When one of the optical axis of birefringent domain be parallel or perpendicular to the optical axis of polarizer or analyzer, then no wave can pass through the analyzer and the domain will be appeared in darkness.

Based on the information provided by AFM, we can conclude that the stripes within these domains were parallel to polarizer or analyzer. The neighboring domains then appeared bright (however, they were not exactly oriented in the diagonal

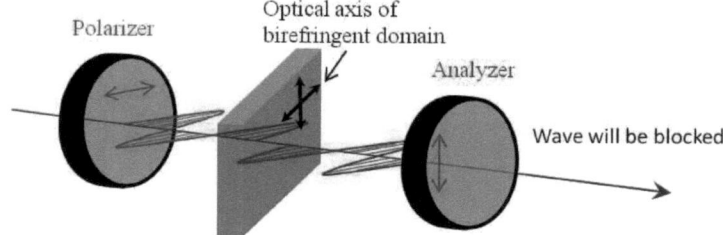

Figure 4.8: A schematic representation of normal orientation for a crystal between crossed polarizers. the arrows indicate the orientation of optical axis for each of the components.

direction). A hexagonal lattice, as revealed by the scattering experiments for our PBLG crystals, corresponds to an optically uniaxial domain, with the optical axis parallel to the **c** axis of the lattice, i.e. along the PBLG helix. The birefringence of PBLG is positive [104, 105] meaning that the optical axis corresponds to the slow axis.

Using a λ plate with its slow axis oriented diagonal to the crossed polarizers, the orientation of the optical axes of the domains, and thus the orientation of the PBLG helices within these domains, can be identified [106]. In absence of any birefringent crystals, a λ plate causes a retardation of 550 nm when it is placed between crossed polarizers. The interference colors expected for different retardations (e.g., resulting from crystals of different thicknesses) are shown in Fig. 4.7f. Fig. 4.7c shows that the bright domains of Fig. 4.7b (see Fig. 4.9a as well) appear in orange color when the corresponding stripes are perpendicular to the slow axis of the λ plate (see Fig. 4.9b as well). This implies that the retardations resulting from the domain structure and the compensator subtract, resulting in a retardation of $\Gamma_1 = 460 \pm 20$ nm. Correspondingly, in Fig. 4.7d the domains appear in blue color, when the slow axis of the compensator is oriented parallel to the stripes (summing up of the retardations, see Fig. 4.7e), resulting in a retardation of $\Gamma_2 = 640 \pm 20$ nm (see Fig. 4.9c as well).

Calculating the differences to the retardation of the λ plate shows that the retardation caused by PBLG crystalline domains is $\Gamma = 90 \pm 20$ nm. This retardation is equal to the product of birefringence $\triangle n$ and the thickness h_c of the crystal: $\Gamma = \triangle n \, h_c$. Knowing the thickness of the crystal, $h_c = 600 \pm 50$ nm from AFM measurement, one can calculate the birefringence of the crystalline domains:

4.5 Birefringent Domains and Molecular Orientation

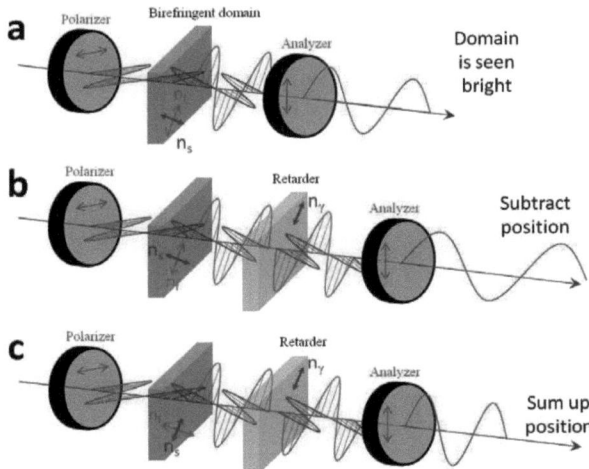

Figure 4.9: A schematic representation for different orientation of Birefringent domain between crossed polarizers and with respect to retarder. (a) When the optical axis of birefringent domain make an angle of 45° with respect to the crossed polarizers, the domain appear bright due to constructive combination of waves passing through these optical axis (see Fig. 4.7b as well). (b, c) Adding a retarder causes a phase difference between two components of light coming from birefringent domain. When optical axis of retarder and slow optical axis of birefringent domain are perpendicular, combination of these two (subtraction of them) will decrease the birefringence (see Fig. 4.7c as well). And when they are parallel, combination of these two (summing up of them) will increase the birefringence (see Fig. 4.7d as well).

$\triangle n = \Gamma/h_c = 0.15 \pm 0.04$. This value is quite different from reported values [104, 105] for the birefringence of PBLG obtained from measurements in liquid crystalline states ($\triangle n = 0.025$). The reason could be that the reported values were obtained from cholesteric liquid crystals containing solvent molecules, which could affect the birefringence properties by swelling the side groups of the PBLG molecules. In addition, the orientational order in the liquid crystalline state is lower than in the crystal, resulting in a smaller birefringence. In a recent study [107], a birefringence of up to 0.091 has been obtained for electrospun fibers of PBLG, much closer to the value obtained by us. However, X-ray experiments revealed that the orientational order of the PBLG molecules in these fibers was still far from being perfect which

65

may result in a smaller birefringence. Hence, the value of 0.15 for our crystals seems to be reasonable, assuming a higher orientational order. After rotation of the crystal by ca. 33°, the same compensator experiments were done for those domains which were dark in Fig. 4.7b. As the direction of the stripes **s**, in these domains was either parallel or perpendicular to the slow axis of the λ plate, we arrived at the same results for the orientation of optical axes (and hence the axis of the α-helices): The slow axis is parallel to the stripes in the domains. Thus, the **c** axis of the PBLG α-helices is also parallel to these stripes (see the vectors **c** and **s** in Fig. 4.7f).

4.6 Zigzag Pattern Formation

It is instructive to compare the observed zig-zag pattern with similar patterns found for example in dried suspensions of the tobacco mosaic virus, another type of rod-like particles, on a substrate [108]. In this study, sharp kinks were observed at the border between adjacent regions of the zig-zag pattern. However, a clear explanation for the origin of this pattern formation was not given. Another example of zig-zag pattern was reported for PBLG liquid crystalline phases. Livolant et. al. [95, 109] identified conditions under which hexagonal columnar phases of PBLG or DNA molecules showed undulating patterns. When increasing the concentration, the undulation patterns were transformed into a herring-bone texture (zig-zag pattern), often observed in hexagonal columnar phases. Further increasing the concentration of macromolecules within the columnar hexagonal phase resulted in additional domains within the previously formed domains. The authors assumed that the undulation textures were expressing the chirality of the molecules. In particular for transitions from the cholesteric to the hexagonal columnar phase, the antagonism between the helically twisted director field of the cholesteric state and the unidirectional order of the hexagonal columnar state was assumed to cause the undulations.

Similar textures of alternating domains have also been observed at the edge (i.e., close to the contact line) of drying drops of lyotropic liquid crystalline phases of DNA [110]. The formation of such patterns was attributed to radial stresses generated during the drying process. Stresses caused undulations which started at the periphery and propagated towards the center of the drop.

Swelling of the side groups of PBLG molecules by solvent [22, 34], or even slight swelling by non-solvent molecules like methanol [21], causes an increase of the separation distance between the PBLG helices. It is much less likely that the presence

4.6 Zigzag Pattern Formation

of some solvent molecules will cause an increase of the length of the α-helix, as this is determined mainly by comparatively strong hydrogen bonds in the core of the molecule. Thus, we do not expect that the length of the helix changes during swelling or de-swelling (drying).

A crystallosolvate phase [34] with a fixed amount of solvent molecules per PBLG monomer can coexist with a dry, crystalline phase in a biphasic region of the phase diagram. Drying of the crystallosolvate phase would mean that the fraction of the crystalline phase increases continuously in the course of the phase transition, accompanied by a corresponding discontinuous decrease of the packing distance in this drying region. On the other hand, Yen et al. reported the existence of a hexagonal columnar liquid crystalline phase in solutions of PBLG in m-cresol [22]. In that phase, the separation distance of the PBLG rods was found to decrease continuously with increasing polymer concentration up to a dry state.

The process of zig-zag pattern formation studied by Livolant et al. [109] differs from that of our experiments, because in our case we followed a transition from the isotropic, and not from the cholesteric, state to a highly ordered state. Hence, we did not start from helically twisted director fields. However, the formation of zig-zag patterns can also be observed in thermotropic [111] and lyotropic [112] hexagonal columnar liquid crystalline phases of non-chiral materials. Theoretically, it was described by Oswald et al. [112] as an undulation instability of the hexagonal columnar phase exposed to a dilative strain perpendicular to the columns, very similar to the undulation instability of smectic layers exposed to a dilative strain along the layer normal [113,114]. The sample dimension perpendicular to the columns (the distance between two glass plates) was considered to be constant, with fixed in-plane column orientation at the boundaries. A net dilative strain upon cooling was discussed to originate from a decrease of the temperature-dependent equilibrium value of the packing distance in the hexagonal lattice, either due to a phase transition or to the positive thermal expansion coefficient. On a short time scale, the liquid crystalline system was able to relax by a bend deformation of the director field. This was much easier than plastic deformation caused by edge dislocations in the two-dimensional hexagonal lattice. Correspondingly, at a (small) critical value of the strain, undulations with a wavelength proportional to the square root of the sample thickness were found to develop. For larger strains, the formation of large amplitude zig-zag patterns was observed, with essentially the same thickness dependence of the period of the pattern as for the undulation wavelength.

If we assume that the here observed crystals, obtained by nucleation and growth, were initially exhibiting a hexagonal columnar liquid crystalline state when embedded in the solution film, one may expect a continuous decrease of the packing distance during the process of drying. We note that in our case the thickness of the objects and the column orientation were not fixed by outer boundaries as in the system of Oswald et al. [112] However, due to gradients in concentration, similar constraints might exist. In particular, one can imagine that drying might lead to thin, solid-like layers at the surfaces of the objects, which could provide appropriate boundary conditions. Further drying of the interior then could explain a net dilative strain larger than the critical value, leading to the observed patterns. Interestingly, as shown in Fig. 4.3f, a power law with an exponent $\beta = 0.52 \pm 0.06$ can be fitted to the dependence of the domain width (corresponding to half of the period of the zig-zag pattern) on the thickness of the objects, in analogy to the square root law for the thickness dependence of the wavelength obtained by Oswald et al. Probably the drying process is too fast to allow for efficient relaxation corresponding to plastic deformation by edge dislocations. Instead of this, the solid-like domains finally split into stripe-like subdomains which are clearly visible in the AFM images.

4.7 Conclusion

In summary, we have shown that large scale PBLG objects grown from semi-dilute thin film solutions show birefringeny. Before drying, these birefringent objects show a single domain structure and are assumed to be in a hexagonal columnar liquid crystalline state. After drying, this single domain splits up into multiple almost parallel domains, each consisting of perfectly aligned parallel stripes. Between adjacent domains, the orientation of the stripes conjointly changed by a certain kink angle with respect to each other which can vary within one crystal by $\pm 4°$ to $5°$. The kink angle averaged over all crystals is $35° \pm 15°$. Scattering experiments indicate a pseudo-hexagonal internal order within these domains, quite consistent with previous observations for bulk samples. Furthermore, we have correlated the optical anisotropy of the objects with the orientation of the PBLG α-helix axis. The helix axis of PBLG was found to be parallel to the stripes of each domain. The formation of zig-zag patterns, often observed in hexagonal columnar liquid crystals, is proposed to be a result of a mechanical instability, caused by the increase of the lateral packing density during the process of drying, and a corresponding net dilative

4.7 Conclusion

strain perpendicular to the columns. In conclusion, combining direct observation of structure formation with characterization by scattering and optical anisotropy experiments allows to follow and to interpret ordering processes in thin film solution of helical polypeptides.

5 General Conclusions and Perspectives

The goal of this work was to study and control the processes of nucleation and growth of poly(γ-benzyl L-glutamate) liquid crystalline and crystalline objects in thin film solutions. Transforming thin solid polymer films into isotropic solutions, via exposure to solvent vapor, allowed us to study in real time, under the optical microscope, the nucleation and growth of ordered solid structures in such solutions. Ordered structures could nucleate in a isotropic polymer solution only when the interactions between the polymer molecules were significantly strong. In dilute solutions below equilibrium volume fraction φ_e, these interactions were not strong enough and did not lead to nucleation. Adding methanol to the isotropic polymer solution leads to a decrease of φ_e below the polymer volume fraction φ_p. Under this condition the interaction forces between the polymer molecules are strong enough to keep many of them together to form a nucleus. These nuclei grow and form hexagonal columnar liquid crystalline structures at large length scales. Removing the methanol from the thin film solution increases φ_e above φ_p and thus dissolution of PBLG objects take place.

Thus, at a constant φ_p, by removing methanol from the film solution, liquid crystalline objects could be dissolved. Interestingly, PBLG objects reformed at the same φ_p when methanol molecules were added again. Thus, by changing the methanol content in the thin film solution, the process of nucleation, growth and dissolution of PBLG objects could be controlled in a reversible fashion. Moreover, adding methanol in a controlled way allowed us to control the supersaturation ratio in the thin film solution and hence to control nucleation rate, number density and growth rate of these objects in real time. Additionally, the variation of the number density of nuclei with the supersaturation ratio for various equilibrium concentrations was found to fit well with predictions of the classical nucleation theory.

We determined how the nucleation density N changed with supersaturation. Analyzing $N(\varphi_\mathrm{p})$, we concluded that the interfacial tension between the ordered structures and the solution increased with the amount of methanol in the thin film solution while decreasing the equilibrium volume fraction. The observation that the nonsolvent influences the equilibrium volume fraction can be explained in two different ways: Calculations on ternary systems of PBLG, solvent and nonsolvent on the basis of Flory's theory show that the equilibrium volume fraction of PBLG in the isotropic phase can strongly decrease with increasing the volume fraction of the nonsolvent and increasing the interaction parameter between polymer and nonsolvent. The other explanation is based on the assumption of complexes between polypeptide molecules and protic nonsolvent molecules like water or methanol leading to different molecules with different solubility properties. In order to check for possible complexation between a protic nonsolvent (methanol in this case) and PBLG molecules, we performed additional NMR measurements which unfortunately gave no evidence for a complexation between PBLG and methanol molecules in the isotropic phase, but still there is no proof against such a complexation in the anisotropic phase of the system. However, NMR experiments confirmed phase separation due to addition of methanol to the system. The drastic decrease of the equilibrium volume fraction with increasing content of methanol, however, can be well understood on the basis of Flory's theory as the effect of a nonsolvent component.

In solution, the formed birefringent objects showed a single domain structure and are assumed to be in a hexagonal columnar liquid crystalline state. After growing these PBLG objects to large scales, they were dried by fast evaporation of solvent and nonsolvent. After drying, the initially single domain liquid crystalline objects split up into multiple almost parallel domains, each consisting of perfectly aligned parallel stripes exhibiting in a zig-zag pattern on the surface of the dried crystals. Between adjacent domains, the orientation of the stripes conjointly changed by a certain kink angle with respect to each other which can vary within one crystal by $\pm\ 4°$ to $5°$. The kink angle averaged over all crystals is $35 \pm 15°$.

In order to find out the internal structure of such highly ordered birefringent objects we performed scattering experiments. X-ray scattering and electron diffraction measurements indicate a pseudo-hexagonal internal order within these domains, quite consistent with previous observations for bulk samples. Furthermore, we have correlated the optical anisotropy of the objects with the orientation of the PBLG α-helix axis. The helix axis of PBLG was found to be parallel to the stripes of each domain.

General Conclusions and Perspectives

At the end of this work, the formation of zig-zag patterns, often observed in hexagonal columnar liquid crystals, is proposed to be a result of a mechanical instability, caused by the increase of the lateral packing density during the process of drying, and a corresponding net dilative strain perpendicular to the columns. In conclusion, combining direct observation of structure formation with characterization by scattering and optical anisotropy experiments allows to follow and to interpret ordering processes in thin film solution of helical polypeptides.

The results of this work can be used in order to understand more generally the nucleation and growth process of crystalline and liquid crystalline structures in biological systems.

In future, it would also be interesting to study the interaction of methanol or other protic nonsolvents with similar poly glutamates like poly(γ-methyl L-glutamate) (PMLG) and poly(γ-ethyl L-glutamate) (PELG) with sufficiently hydrophilic side chains instead of hydrophobic benzyl side chain in PBLG. As the side chains of those polymers are hydrophilic, it would be more plausible to observe interaction or complexation between protic nonsolvents and these side chains in solution by performing NMR measurements. The results can be compared with ours in order to propose a mechanism for changing the equilibrium volume fraction in solutions of these polymers. Then, changing the interfacial tension between the formed objects and the surrounding isotropic phase could be explained based on such complexations. So that, a mechanisms that drive peptide self-assembly in solution and at interfaces can be proposed.

In the next step one important question would be that how we can functionalize these structures as nanomaterials. The potential applications of these peptidic crystalline materials in nanotechnology, medicine, etc. could be assessed.

At the end, we would like to mention another interesting feature of this helical polypeptides. they possess a huge overall dipole moment which could interact with external electric field. It would be interesting to use this feature and study the effect of electric field on the growth process of objects in order to develop systems with directed assembly which can be used for example in biological systems as a dynamic molecular system with customized morphological response.

Bibliography

[1] Whitesides, G. M.; Mathias, J. P.; Seto, C. T. *Science* **1991**, *254*, 1312-1319.

[2] Vauthey, S.; Santoso, S.; Gong, H.; Watson, N.; Zhang, S. *Proc. Natl. Acad. Sci. USA* **2002**, *99*, 5355-5360.

[3] Palmer, L. C.; Stupp, S. I. *Accounts of Chemical Research* **2008**, *41*, 1674-1684.

[4] Floudas, G.; Papadopoulos, P.; Klok, H. A.; Vandermeulen, G. W. M.; Rodriguez-Hernandez, J. *Macromolecules* **2003**, *36*, 3673-3683.

[5] Bawden, F. C.; Pirie, N. W. *Nature* **1938**, *141*, 513-514.

[6] Bernal, J. D.; Fankuchen, I. *J. Gen. Physiol.* **1941**, *25*, 111-146.

[7] Oster, G. *J. Gen. Physiol.* **1950**, *33*, 445-473.

[8] Miller, W. G. *Annu. Rev. Phys. Chem.* **1978**, *29*, 519-535.

[9] Snyder, J. A.; McIntosh, J. R. *Annu. Rev. Biochem.* **1976**, *45*, 699-720.

[10] Elliott, A.; Ambrose, E. J. *Discuss. Faraday Soc.* **1950**, *9*, 246-251.

[11] Doty, P.; Bradbury, J. H.; Holtzer, A. M. *J. Am. Chem. Soc.* **1956**, *78*, 947-954.

[12] Perutz, M. F. *Nature* **1951**, *167*, 1053-1054.

[13] Robinson, C. *Trans. Farad. Soc.* **1956**, *52*, 571–592.

[14] Cohen, Y. *Sructure Formation in Solutions of Rigid Polymers Undergoing c Phase Transition;* PhD thesis: The University of Massachusetts, 1987.

[15] Leforestier, A.; Livolant, F. *Biol Cell* **1991**, *71*, 115-122.

[16] Luzzati, V.; Cesari, M.; Spach, G.; Masson, F.; Vincent, J. M. *J. Mol. Biol.* **1961**, *3*, 566-584.

[17] Miller, W.; Wu, C. C.; Wee, E. L.; Santee, G. L.; Rai, J. H.; Goebel, K. G. *Pure Appl. Chem.* **1974**, *38*, 37-58.

[18] Flory, P. J. *Proc. R. Soc. Lond. A* **1956**, *234*, 73–89.

[19] Flory, P. J. Molecular Theory of Liquid Crystals. In *Liquid Crystal Polymers I*, Vol. 59; Gordon, M.; Platé, N. A., Eds.; Springer: 1984.

[20] Parry, D. A. D.; Elliott, A. *J. Mol. Biol.* **1967**, *25*, 1-13.

[21] Watanabe, J.; Imai, K.; Gehani, R.; Uematsu, I. *J. Polym. Sci. Pol. Phys.* **1981**, *19*, 653–665.

[22] Yen, C.-C.; Edo, S.; Oka, H.; Tokita, M.; Watanabe, J. *Macromolecules* **2008**, *41*, 3727–3733.

[23] Russo, P. S.; Miller, W. G. *Macromolecules* **1984**, *17*, 1324-1331.

[24] Rybnikar, F.; Geil, P. H. *Biopolym.* **1972**, *11*, 271-278.

[25] Blais, J. J. B. P.; Geil, P. H. *J. Ultrastruct. Res.* **1968**, *22*, 303-311.

[26] Nelson, D. L.; Cox, M. M. *Lehninger principles of biochemistry;* W. H. Freeman: New York, 4 ed.; 2004.

[27] Lehninger, A. L.; Nelson, D. L.; Cox, M. M. *principles of biochemistry;* Worth Publishers: New York, 3 ed.; 2000.

[28] Botiz, I. *Processes of Ordered Structure Formation in Polypeptide Thin Film Solutions;* PhD thesis: Université de Haute Alsace Mulhouse, 2007.

[29] Pauling, L.; Corey, R. B.; Branson, H. R. *Proc. Natl. Acad. Sci. USA.* **1951**, *37*, 205–211.

[30] Bamford, C. H.; Hanby, W. E.; Happey, F. *Proc. R. Soc. Lond. A* **1951**, *205*, 30-47.

[31] Doty, P.; Holtzer, A. M.; Bradbury, J. H.; Blout, E. R. *J. Am. Chem. Soc.* **1954**, *76*, 4493-4494.

Bibliography

[32] Doty, P.; Yang, J. T. *J. Am. Chem. Soc.* **1956**, *78*, 498–500.

[33] Moffitt, W. *P. Natl. Acad. Sci. USA* **1956**, *42*, 736–746.

[34] Ginzburg, B. M.; Shepelevskii, A. A. *Journal of Macromolecular Science B* **2003**, *42*, 1-56.

[35] Zimmel, J. M.; Wu, C. C.; Miller, W. G.; Mason, R. P. *J. Phys. Chem.* **1983**, *87*, 5435-5443.

[36] Dietz, M. *Helix - Coil - Transition in Solid Polypeptides;* PhD thesis: Minz, 2007.

[37] McKinnon, A. J.; Tobolsky, A. *J. Phys. Chem.* **1968**, *72*, 1157–1161.

[38] Wada, A. *J. Chem. Phys.* **1959**, *30*, 328-329.

[39] Papkov, S. P. Liquid Crystalline Order in Solutions of Rigid-Chain Polymers. In *Liquid Crystal Polymers I*, Vol. 59; Gordon, M.; Platé, N. A., Eds.; 1984.

[40] Ciferri, A. *Developments in Oriented Polymers;* volume 2 Elsevier Applied Sci.: England: London, New York, 1987.

[41] Tsutsui, T.; Tanaka, R.; Tanaka, T. *J. Polym. Sci.: Polym. Lett. Ed.* **1979**, *17*, 511-520.

[42] Nakajima, A.; Kugo, K.; Hayashi, T. *Polym. J.* **1979**, *11*, 995-1001.

[43] Sumimoto, H.; Hashimoto, K. *Adv. Polym. Sci.* **1985**, *64*, 63-91.

[44] Bouligand, Y. *Liquid Crystalline Order in Polymers;* Academic Press: New York, 1978.

[45] Russo, P. S.; Miller, W. G. *Macromolecules* **1983**, *16*, 1690-1693.

[46] Ginzburg, B.; Syromyatnikova, T.; Frenkel, S. *Polymer Bulletin* **1985**, *13*, 139-144.

[47] Sasaki, S.; Hikata, M.; Shiraki, C.; Uematsu, I. *Polym. J.* **1982**, *14*, 205-213.

[48] Nakajima, A.; Hayashi, T.; Ohmori, M. *Biopolymers* **1968**, *6*, 973-982.

[49] Pethrick, R. A. *Polymer Structure Characterisation: From Nano to Macro Organization;* The Royal Society of Chemistry: Thomas Graham House, Science Park, Milton Road, Cambridge CB4 0WF, UK, 2007.

[50] Erdemir, D.; Lee, A. Y.; Myerson, A. S. *Acc. Chem. Res.* **2009**, *42*, 621-629.

[51] Mohanty, R.; Bhandarkar, S.; Estrin, J. *AIChE Journal* **1990**, *36*, 1536-1544.

[52] Nielsen, A. E.; Söhnel, O. *Journal of Crystal Growth* **1971**, *11*, 233-242.

[53] Arora, D. *Structure-property Evolution During Polymer Crystallization;* PhD thesis: Amherst, 2010.

[54] De Yoreo, J. J.; Vekilov, P. G. *Principles of crystal nucleation and growth. In: Biomineralization;* volume 54 Mineralogical Society of America: Washington, D.C., 2003.

[55] Roelands, C. P. M.; Horst, J. H. t.; Kramer, H. J. M.; Jansens, P. J. *Crystal Growth and Design* **2006**, *6*, 1380-1392.

[56] Mersmann, A. *Journal of Crystal Growth* **1990**, *102*, 841-847.

[57] Söhnel, O. *Journal of Crystal Growth* **1982**, *57*, 101-108.

[58] Chowdhury, M. *Thin Polymer Films Out of Thermodynamic Equilibrium;* PhD thesis: Universität Freiburg, 2012.

[59] Rahimi Harchegani, K. *Controlling the Crystal Growth and Morphology of Conjugated Polymers;* PhD thesis: Universität Freiburg, 2013.

[60] Lin, Z.; Kerle, T.; Baker, S. M.; Hoagland, D. A.; Schäffer, E.; Steiner, U.; Russell, T. P. *J. Chem. Phys.* **2001**, *114*, 2377-2381.

[61] Botiz, I.; Grozev, N.; Schlaad, H.; Reiter, G. *Soft Matter* **2008**, *4*, 993-1002.

[62] Hermes, F. J. *Polypeptide-Hybrid Block Copolymers: Chain Length and Conformation Effects on the Self-Assembly in Solution;* PhD thesis: Universität Potsdam, 2010.

[63] Schäffer, E. *Instabilities in Thin Polymer Films: Structure Formation and Pattern Transfer;* PhD thesis: Universität Konstanz, 2001.

[64] Dierking, I. *Textures of Liquid Crystals;* WILEY-VCH Verlag GmbH and Co. KGaA: Weinheim, 2003.

[65] Bhushan, B. *Nanotribology and Nanomechanics;* Springer-Verlag Berlin: Heidelberg, 2 ed.; 2008.

[66] Kaupp, G. *Atomic force microscopy, Scanning Nearfield, Optical Microscopy and Nanoscratching: Application to Rough and Natural Surfaces;* Springer-Verlag Berlin: Heidelberg, 2006.

[67] Shokri, R. *Self-Assembly of Supra-Molecular Systems on Graphene or Graphite;* PhD thesis: Universität Freiburg, 2013.

[68] *JPK NanoWizard Handbook;* 2005.

[69] Morita, S.; Giessibl, F. J.; Wiesendanger, R. *Noncontact Atomic Force Microscopy: Volume 2;* Springer-Verlag Berlin: Heidelberg, 2009.

[70] Weisenhorn, A. L.; Hansma, P. K.; Albrecht, T. R.; Quate, C. F. *Appl. Phys. Lett.* **1989**, *54*, 2651-2653.

[71] Martin, Y.; Williams, C. C.; Wickramasinghe, H. K. *J. Appl. Phys.* **1987**, *61*, 4723-4729.

[72] Giessibl, F. J. *Rev. Mod. Phys.* **2003**, *75*, 949-983.

[73] Jenkins, R. *X-ray Techniques: Overview. in Encyclopedia of Analytical Chemistry;* John Wiley and Sons Ltd: Chichester, 2000.

[74] Vinci, R.; Zielinski, E.; Bravman, J. *Thin Solid Films* **1995**, *262*, 142-153.

[75] Rachwal, J. D. *X-Ray Diffraction Applications in Thin Films and (100) Silicon Substrate Stress Analysis;* PhD thesis: University of South Florida, 2010.

[76] Callister, W. D. J. *Materials Science and Engineering: An Introduction;* John Wiley and Sons, Inc.: Utah, 7 ed.; 2007.

[77] Kittel, C. *Introduction to Solid State Physics;* John Wiley and Sons, Inc: Berkeley, 8 ed.; 2005.

[78] Bozzola, J. J.; Russel, L. D. *Electron Microscopy: Principles and Techniques for Biologists;* Jones and Barlett Publishers International: London, 2 ed.; 1999.

[79] Kleber, W. *Einführung in die Kristallographie;* VEB Verlag Technik: Berlin, 1982.

[80] Kristensen, M. *Wholegrains and Dietary Fibres: Impact on Body Weight, Appetite Regulation and Nutrient Digestibility;* PhD thesis: University of Copenhagen, 2009.

[81] Lambert, J. B.; Mazzola, E. P. *Nuclear Magnetic Resonance 5pectroscopy: An Introduction to Principles, Applications, and Experimental Methods;* Prentice Hall: NewJersey, 2004.

[82] Berg, G. M.; Tymoczko, J. L.; Stryer, L. *Biochemistry;* W. H. Freeman: New York, 5 ed.; 2002.

[83] Pavia, D. L.; Lampman, G. M.; Kriz, G. S.; Vyvyan, J. R. *Introduction to Spectroscopy;* Brooks/Cole Cengage Learning: Washington, 4 ed.; 2009.

[84] Dunn, W. B.; Ellis, D. I. *Trends in Analytical Chemistry* **2005**, *24*, 285-294.

[85] Trolier-McKinstry, S.; Koh, J. *Thin Solid Films* **1998**, *313-314*, 389-393.

[86] McMarr, P. J.; Vedam, K.; Narayan, J. *J. Appl. Phys.* **1986**, *59*, 694-701.

[87] Smith, T. *Surface Science* **1976**, *56*, 252-271.

[88] Pellegrino, L.; Canepa, M.; Gonella, G.; Bellingeri, E.; Marre, D.; Tumino, A.; Siri, A. *J. Phys. IV France* **2001**, *11*, 337-342.

[89] Crossland, E.; Rahimi, K.; Reiter, G.; Steiner, U.; Ludwigs, S. *Adv. Funct. Mater.* **2011**, *21*, 518-524.

[90] Onsager, L. *Ann. NY Acad. Sci.* **1949**, *51*, 627–659.

[91] Yu, S. M.; Conticello, V. P.; Zhang, G.; Kayser, C.; Fournier, M. J.; Mason, T. L.; Tyrell, D. A. *Nature* **1997**, *389*, 167-170.

[92] Yen, C.-C.; Taguchi, Y.; Tokita, M.; Watanabe, J. *Mol. Cryst. Liq. Cryst.* **2010**, *516*, 91–98.

[93] Terentjev, E. M.; Osipov, M. A.; Sluckin, T. J. *J. Phys. A: Math. Gen.* **1994**, *27*, 7047-7059.

Bibliography

[94] de Gennes, P. G.; Prost, J. *The Physics of Liquid Crystals;* Clarendon Press: 2nd ed.; 1993.

[95] Livolant, F.; Leforestier, A. *Prog. Polym. Sci.* **1996**, *21*, 1115-1164.

[96] Botiz, I.; Schlaad, H.; Reiter, G. *Adv. Polym. Sci.: Self-Organized Nanostructures of Amphiphilic Block Copolymers;* volume 242 Springer Berlin-Heidelberg: Berlin-Heidelberg, 2011.

[97] *SNE Research Co. LTD.* www.solarnenergy.com.

[98] Imai, H.; Oaki, Y. *MRS Bulletin* **2010**, *35*, 138-144.

[99] Wee, E. L.; Miller, W. G. *J. Phys. Chem.* **1971**, *75*, 1446-1452.

[100] Malcolm, B. R. *J. Polym. Sci.: Part C* **1971**, *34*, 87-99.

[101] Robinson, C.; Ward, J. C.; Beevers, R. B. *Discuss. Faraday Soc.* **1958**, *25*, 29-42.

[102] McKinnon, A. J.; Tobolsky, A. *J. Phys. Chem.* **1966**, *70*, 1453–1456.

[103] Ludwigs, S.; Krausch, G.; Reiter, G.; Losik, M.; Antonietti, M.; Schlaad, H. *Macromolecules* **2005**, *38*, 7532-7535.

[104] Du Pré, D. B.; Samulski, E. T. Polypeptide Liquid Crystals. In *Liquid Crystals, the Fourth State of Matter*; Saeva, F. D., Ed.; Marcel Dekker, Inc.: 1979.

[105] Robinson, C.; Ward, J. C. *Nature* **1957**, *180*, 1183–1184.

[106] Sawyer, L.; Grubb, D.; Meyers, G. F. *Polymer Microscopy;* Springer: New York, 3 ed.; 2008.

[107] Tsuboi, K.; Marcelletti, E.; Matsumoto, H.; Ashizawa, M.; Minagawa, M.; Furuya, H.; Tanioka, A.; Abe, A. *Polymer Journal* **2012**, *44*, 360-365.

[108] Maeda, H. *Langmuir* **1997**, *13*, 4150-4161.

[109] Livolant, F.; Bouligand, Y. *J. Physique* **1986**, *47*, 1813-1827.

[110] Smalyukh, I. I.; Zribi, O. V.; Butler, J. C.; Lavrentovich, O. D.; Wong, G. C. L. *Phys. Rev. Lett.* **2006**, *96*, 177801-177804.

[111] Gharbia, M.; Cagnon, M.; Durand, G. *J. Physique Lett.* **1985**, *46*, L-683–L-687.

[112] Oswald, P.; Géminard, J. C.; Lejcek, L.; Sallen, L. *J. Phys. II France* **1996**, *6*, 281-303.

[113] Clark, N. A.; Meyer, R. B. *Appl. Phys. Lett.* **1973**, *22*, 493-494.

[114] Delaye, M.; Ribotta, R.; Durand, G. *Physics Letters A* **1973**, *44*, 139-140.

[115] Jennings, B. R.; Jerrard, H. G. *Nature* **1966**, *210*, 90.

[116] Papadopoulos, P.; Floudas, G.; Schnell, I.; Klok, H.-A.; Aliferis, T.; Iatrou, H.; Hadjichristidis, N. *J. Chem. Phys.* **2005**, *122*, 224906-1–224906-4.

[117] Lavigne, P.; Tancrède, P.; Lamarche, F.; Grandbois, M.; Salesse, C. *Thin Solid Films* **1994**, *242*, 229-233.

[118] Doi, M.; Edwards, S. F. *The Theory of Polymer Dynamics;* Clarendon Press: Oxford, 1986.

[119] Khokhlov, A. R. Theories Based on the Onsager Approach. In *Liquid Crystallinity in Polymers: Principles and Fundamental Properties*; Ciferri, A., Ed.; VCH Publishers: 1991.

[120] Burgers, J. M. *Verh. Kon. Ned. Akad. Wet.* **1938**, *16*, 113.

[121] Boersma, S. *J. Chem. Phys.* **1960**, *32*, 1626-1631.

[122] Wada, A. *J. Chem. Phys.* **1959**, *30*, 329-330.

Appendix A

PBLG Specification

Chemical sum formula:

$$[C_{12}H_{13}NO_3]_n \qquad (A.1)$$

molar mass of monomer7

$$M_m = 219.237 \, \text{g mol}^{-1} \qquad (A.2)$$

translation of monomeric length increment along the α-helical axis [115]

$$L_{\text{monomeric}} = 0.15 \, \text{nm} \qquad (A.3)$$

specific volume v of α-helix at atmospheric pressure [116]:

$$v_p(T) = A_0 + A_1 T + A_2 T^2$$

$$= \left[0.788 + 4.92 \cdot 10^{-4} \cdot \frac{T}{^\circ\text{C}} + 7.57 \cdot 10^{-7} \cdot \left(\frac{T}{^\circ\text{C}}\right)^2 \right] \text{cm}^3 \, \text{g}^{-1} \qquad (A.4)$$

density at 0 °C:

$$\rho_p = \frac{1}{v} = 1.269 \, \text{g cm}^{-3} \qquad (A.5)$$

volume per monomer at $0\,^\circ\mathrm{C}$:

$$V_\mathrm{m} = \frac{v_\mathrm{p} M_\mathrm{m}}{N_\mathrm{A}} = \frac{0.788 \cdot 10^{-6}\,\mathrm{m^3\,g^{-1}} \cdot 219.237\,\mathrm{g\,mol^{-1}}}{6.022 \cdot 10^{23}\,\mathrm{mol^{-1}}} = 2.869 \cdot 10^{-28}\,\mathrm{m^3} \quad (A.6)$$

area contribution per helix at $0\,^\circ\mathrm{C}$:

$$A = \frac{V_\mathrm{m}}{L_\mathrm{monomeric}} = \frac{2.869 \cdot 10^{-28}\,\mathrm{m^3}}{0.15 \cdot 10^{-9}\,\mathrm{m}} = 1.912 \cdot 10^{-18}\,\mathrm{m^2} \quad (A.7)$$

lattice constant of a hexagonal packing of α-helix rods:

$$a = \sqrt{\frac{A}{\sin\gamma}} = \sqrt{\frac{1.912 \cdot 10^{-18}\,\mathrm{m^2}}{\sin 120^\circ}} = 1.486 \cdot 10^{-9}\,\mathrm{m} \quad (A.8)$$

corresponds well with pseudo-hexagonal lattice with $a = b = 1.48\,\mathrm{nm}$ *(form C)* [92]. Monomer volume also corresponds well with monoclinic lattice with $a = 2.9\,\mathrm{nm}$, $b = 1.34\,\mathrm{nm}$, $c = 2.69\,\mathrm{nm}$, $\alpha = \gamma = 90^\circ$, $\beta = 96^\circ$ *(form B)* [92], assuming $c = 18 L_\mathrm{monomeric}$ (18/5 helix) and centered monoclinic lattice ($2 \cdot 18$ monomers per unit cell):

$$V_\mathrm{m} = \frac{abc\sin\beta}{2} = \frac{2.9 \cdot 10^{-9}\,\mathrm{m} \cdot 1.34 \cdot 10^{-9}\,\mathrm{m} \cdot 2.69 \cdot 10^{-9}\,\mathrm{m} \cdot \sin 96^\circ}{2 \cdot 18} = 2.888 \cdot 10^{-28}\,\mathrm{m^3} \quad (A.9)$$

Dipole moment as vector sum of monomer contributions in a 18/5 helix with degree of polymerization n:

$$\boldsymbol{\mu}(n) = \boldsymbol{\mu}_\parallel(n) + \boldsymbol{\mu}_\perp(n) = n\boldsymbol{\mu}_{\parallel,\mathrm{m}} + \mu_{\perp,\mathrm{m}} \sum_{i=0}^{n-1} (\cos\varphi_i, \sin\varphi_i, 0)$$

$$= n\boldsymbol{\mu}_{\parallel,\mathrm{m}} + \mu_{\perp,\mathrm{m}} \sum_{i=0}^{n-1} (\cos(i \cdot 100^\circ), \sin(i \cdot 100^\circ), 0) \quad (A.10)$$

$\boldsymbol{\mu}_\perp(n)$: period of 18, coordinates of an octadecagon. Extrema: $\mu_\perp(18) = 0$, $\mu_\perp(9) = 1.305\,\mu_{\perp,\mathrm{m}}$. Experimental values obtained from surface potential of monolayers on water: $0.07\,\mathrm{D} = 2.33 \cdot 10^{-31}\,\mathrm{C\,m} \leq \mu_\perp \leq 0.23\,\mathrm{D} = 7.66 \cdot 10^{-31}\,\mathrm{C\,m}$ [117]. Spontaneous polarization of a smectic C*-like structure with tilt angle ϑ and number

Appendix A

density ϱ_μ of dipoles:

$$P_s = P_0 \sin \vartheta < \varrho_\mu \mu_\perp (n) \sin \vartheta \qquad (A.11)$$

Birefringence $\Delta n = n_\parallel - n_\perp = 0.025 > 0$ [104, 105].

Concentration Regimes

Number density ϱ_p of polypeptide molecules of length L in dilute solutions [118]:

$$\varrho_p = \frac{N_p}{V} \lesssim \varrho_1 \simeq \frac{1}{L^3} = \frac{1}{n^3 L^3_{\text{monomeric}}} \qquad (A.12)$$

volume fraction

$$\varphi_p = \frac{V_p}{V} = \frac{N_p n V_m}{V} = \varrho_p n V_m \qquad (A.13)$$

with polypeptide volume V_p and number of polypeptide molecules N_p. However, in our experiments we used a different way to identify the PBLG volume fraction in thin film solution.

Weight fraction (in one-component solvent s)

$$w_p = \frac{m_p}{m} = \frac{V_p \rho_p}{V_p \rho_p + V_s \rho_s} = \frac{\varphi_p \rho_p}{\varphi_p \rho_p + (1 - \varphi_p) \rho_s} \qquad (A.14)$$

limiting volume fraction

$$\varphi_1 = \varrho_1 n V_m = \frac{V_m}{n^2 L^3_{\text{monomeric}}} \qquad (A.15)$$

Number density ϱ_p of polypeptide molecules with diameter b for semi dilute solutions [118]:

$$\varrho_1 \lesssim \varrho_p \ll \varrho_2 \simeq \frac{1}{bL^2} = \frac{1}{bn^2 L^2_{\text{monomeric}}} = \varrho_1 \frac{L}{b} = \varrho_1 \frac{n L_{\text{monomeric}}}{b} \qquad (A.16)$$

for $n = 187$:

$$\varphi_1 \approx \frac{2.869 \cdot 10^{-28} \, \text{m}^3}{187^2 \cdot (0.15 \cdot 10^{-9} \, \text{m})^3} = 0.00243 \qquad (A.17)$$

$$\frac{L}{b} \approx \frac{187 \cdot 0.15 \cdot 10^{-9}\,\mathrm{m}}{1.486 \cdot 10^{-9}\,\mathrm{m}} = 18.9 \qquad (A.18)$$

$$\varphi_2 = \varphi_1 \frac{L}{b} \approx 0.0459 \qquad (A.19)$$

in chloroform of density $\rho_\mathrm{s} = 1.48\,\mathrm{g\,cm^{-3}}$:

$$w_1 = \frac{\varphi_1 \rho_\mathrm{p}}{\varphi_1 \rho_\mathrm{p} + (1-\varphi_1)\rho_\mathrm{s}} \approx \varphi_1 \frac{\rho_\mathrm{p}}{\rho_\mathrm{s}} = 0.00243 \cdot \frac{1.269\,\mathrm{g\,cm^{-3}}}{1.48\,\mathrm{g\,cm^{-3}}} = 0.00208 \qquad (A.20)$$

$$w_2 \approx w_1 \frac{L}{b} \approx 0.0394 \qquad (A.21)$$

The Onsager theory [90] results for the isotropic state in an upper volume fraction limit

$$\varphi_2^\mathrm{i} = 3.24 \frac{b}{L} = 0.171 \qquad (A.22)$$

and for the anisotropic state in a lower volume fraction limit

$$\varphi_2^\mathrm{a} = 4.49 \frac{b}{L} = 0.238 \qquad (A.23)$$

and for $L/b = 18.9$. An improved version of the Onsager theory [119], not using the second virial approximation, results in

$$\varphi_2^\mathrm{i} = 3.29 \frac{b}{L} = 0.174 \qquad (A.24)$$

[34]and

$$\varphi_2^\mathrm{a} = 4.191 \frac{b}{L} = 0.222 \qquad (A.25)$$

The Flory theory [18] for an athermal system (Flory-Huggins interaction parameter

Appendix A

$\chi = 0$) results in [119]

$$\varphi_2^i = 7.89\frac{b}{L} = 0.417 \tag{A.26}$$

and

$$\varphi_2^a = 11.57\frac{b}{L} = 0.612 \tag{A.27}$$

Weight fraction of PBLG in DMF for a crystallosolvate with 3 solvent molecules per repeat unit [34]:

$$w_p = \frac{2M_m}{2M_m + 3M_{DMF}} = \frac{1}{1 + 1.5 M_{DMF}/M_m} = \frac{1}{1 + 1.5 \cdot 73.094/219.237} = 0.6666 \tag{A.28}$$

Rotational Diffusion

Rotational friction constant $\zeta_{r0} = \tau/\omega$ (with torque τ and angular velocity ω) of a prolate ellipsoid with half axes a and b and an aspect ratio $p = a/b > 1$ in dilute solutions with a solvent viscosity η_s (see [118] and references therein):

$$\zeta_{r0} = \frac{16\pi}{3}\eta_s a^3 \left(1 - \frac{1}{p^4}\right)\left[\frac{2p^2 - 1}{2p(p^2 - 1)^{1/2}} \ln\left(\frac{p + (p^2 - 1)^{1/2}}{p - (p^2 - 1)^{1/2}}\right) - 1\right]^{-1} \tag{A.29}$$

for $p \gg 1$:

$$\zeta_{r0} \approx \frac{16\pi}{3}\eta_s a^3 \left[\ln\left(\frac{p + p(1 - 1/p^2)^{1/2}}{p - p(1 - 1/p^2)^{1/2}}\right) - 1\right]^{-1}$$

$$\approx \frac{16\pi}{3}\eta_s a^3 \left[\ln\left(\frac{2p - 1/(2p)}{1/(2p)}\right) - 1\right]^{-1} \approx \frac{16\pi \eta_s a^3}{3\left[2\ln(2p) - 1\right]} \tag{A.30}$$

with length $L = 2a$:

$$\zeta_{r0} \approx \frac{\pi \eta_s L^3}{3\left[\ln(L/b) - 1/2\right]} = \frac{\pi \eta_s L^3}{3\left[\ln(L/b) - \gamma\right]} \tag{A.31}$$

with $\gamma = 1/2$ for ellipsoids.

For cylinders of length L and radius b: $\gamma = 0.8$ [120], $\gamma = \gamma(L/b)$ [121].

Rotational diffusion constant

$$D_\mathrm{r} = \frac{k_\mathrm{B} T}{\zeta_\mathrm{r}} \tag{A.32}$$

angular frequency of Debye relaxation process:

$$\omega_\mathrm{r} = 2\pi f_\mathrm{r} = 2 D_\mathrm{r} = \frac{2kT}{\zeta_\mathrm{r}} \tag{A.33}$$

Debye relaxation time:

$$t_\mathrm{r} = \frac{1}{\omega_\mathrm{r}} = \frac{\zeta_\mathrm{r}}{2 k_\mathrm{B} T} \tag{A.34}$$

from Wada [122]:

$$t_\mathrm{r0} = t_0 \frac{2(1-\rho^4)}{\frac{3(2-\rho^2)\rho^2}{(1-\rho^2)^{1/2}} \cdot \ln\left[\frac{1+(1-\rho^2)^{1/2}}{\rho}\right] - 3\rho^2} \tag{A.35}$$

with

$$t_0 = \frac{4\pi a b^2 \eta_\mathrm{s}}{kT} \tag{A.36}$$

and

$$\rho = \frac{b}{a} = \frac{1}{p} \tag{A.37}$$

$$t_\mathrm{r0} = \frac{4\pi a b^2 \eta_\mathrm{s}}{3kT} \cdot \frac{2(1-\rho^4)}{\frac{(2-\rho^2)\rho^2}{(1-\rho^2)^{1/2}} \ln\left[\frac{1+(1-\rho^2)^{1/2}}{\rho}\right] - \rho^2} \tag{A.38}$$

$$\zeta_\mathrm{r0} = 2 k_\mathrm{B} T t_\mathrm{r0} = \frac{8\pi a b^2 \eta_\mathrm{s}}{3} \cdot \frac{2(1-\rho^4)}{\frac{(2-\rho^2)\rho^2}{(1-\rho^2)^{1/2}} \ln\left(\frac{1+(1-\rho^2)^{1/2}}{\rho}\right) - \rho^2}$$

Appendix A

$$= \frac{16\pi}{3} a^3 \eta_s \left(1 - \rho^4\right) \left[\frac{2 - \rho^2}{(1 - \rho^2)^{1/2}} \ln\left(\frac{1 + (1 - \rho^2)^{1/2}}{\rho}\right) - 1\right]^{-1}$$

$$= \frac{16\pi}{3} a^3 \eta_s \left(1 - \frac{1}{p^4}\right) \left[\frac{2p^2 - 1}{p(p^2 - 1)^{1/2}} \ln\left(p + \left(p^2 - 1\right)^{1/2}\right) - 1\right]^{-1} \quad \text{(A.39)}$$

for $p \gg 1$:

$$\zeta_r \approx \frac{16\pi a^3 \eta_s}{3 \left[2 \ln(2p) - 1\right]} = \frac{\pi \eta_s L^3}{3 \left[\ln(L/b) - 1/2\right]} \quad \text{(A.40)}$$

Rotational diffusion constant in semi dilute solution

$$D_r = \beta \frac{D_{r0}}{(\rho_p L^3)^3} \quad \text{(A.41)}$$

with β of the order of 10^3 [118].

Electric Dipole Moment

Electric polarization

$$\mathbf{P} = \mathbf{P}_\alpha + \mathbf{P}_\mu = \varepsilon_0 \chi \mathbf{E} = \varepsilon_0 (\varepsilon - 1) \mathbf{E} \quad \text{(A.42)}$$

$$\mathbf{P}_\mu = \frac{\sum_i \boldsymbol{\mu}_i}{V} \quad \text{(A.43)}$$

contribution by orientation of dipoles in dilute solution with number density ϱ_μ in electric field:

$$P_\mu = \varrho_\mu \mu \langle \cos \vartheta \rangle \quad \text{(A.44)}$$

energy of dipole moment $\boldsymbol{\mu}$ in an electric field $\mathbf{E}$:

$$W(\vartheta) = -\boldsymbol{\mu} \mathbf{E} = -\mu E \cos \vartheta \quad \text{(A.45)}$$

average by Boltzmann statistic:

$$\langle \cos\vartheta \rangle = \frac{\int_0^\pi \cos\vartheta \exp\left(-W(\vartheta)/k_\mathrm{B}T\right)\sin\vartheta\, d\vartheta}{\int_0^\pi \exp\left(-W(\vartheta)/k_\mathrm{B}T\right)\sin\vartheta\, d\vartheta}$$

$$= \frac{\int_0^\pi \cos\vartheta \exp\left(\mu E \cos\vartheta/k_\mathrm{B}T\right)\sin\vartheta\, d\vartheta}{\int_0^\pi \exp\left(\mu E \cos\vartheta/k_\mathrm{B}T\right)\sin\vartheta\, d\vartheta} \tag{A.46}$$

$$x = \frac{\mu E}{kT}, \qquad v = \cos\vartheta, \qquad dv = -\sin\vartheta\, d\vartheta \tag{A.47}$$

$$\langle \cos\vartheta \rangle = \frac{\int_{-1}^{1} v \exp(xv)\, dv}{\int_{-1}^{1} \exp(xv)\, dv} \tag{A.48}$$

$$w = \frac{1}{x}\exp(xv), \qquad w' = \exp(xv), \qquad v' = 1 \tag{A.49}$$

$$\int v \exp(xv)\, dv = \int vw'\, dv = vw - \int v'w\, dv$$

$$= \frac{v}{x}\exp(xv) - \frac{1}{x}\int \exp(xv)\, dv$$

$$= \frac{v}{x}\exp(xv) - \frac{1}{x^2}\exp(xv) = \frac{vx-1}{x^2}\exp(xv) \tag{A.50}$$

$$\langle \cos\vartheta \rangle = \frac{\frac{1}{x^2}\left[(vx-1)\exp(xv)\right]_{v=-1}^{1}}{\frac{1}{x}\left[\exp(xv)\right]_{v=-1}^{1}} = \frac{(x-1)\exp(x) + (x+1)\exp(-x)}{x\left[\exp(x) - \exp(-x)\right]}$$

$$= \frac{x\left[\exp(x) + \exp(-x)\right] - \left[\exp(x) - \exp(-x)\right]}{x\left[\exp(x) - \exp(-x)\right]}$$

Appendix A

$$= \frac{\exp(x) + \exp(-x)}{\exp(x) - \exp(-x)} - \frac{1}{x}$$

$$= \coth x - \frac{1}{x} = \frac{1}{x} + \frac{x}{3} - \frac{x^3}{45} + \ldots - \frac{1}{x} = \frac{x}{3} - \frac{x^3}{45} + \ldots \tag{A.51}$$

for $x \ll 1$:

$$\langle \cos \vartheta \rangle \approx \frac{x}{3} = \frac{\mu E}{3kT} \tag{A.52}$$

fractional error f in $\langle \cos \vartheta \rangle$:

$$f = \frac{x^2}{15} \tag{A.53}$$

with $\mu = 2460\,\mathrm{D} = 8.2 \cdot 10^{-27}\,\mathrm{C\,m}$ for molar mass $M = 154000\,\mathrm{g\,mol^{-1}}$ [38], $T = 300\,\mathrm{K}$, and $E = 10^5\,\mathrm{V/m}$,

$$f = \frac{1}{15}\left(\frac{\mu E}{kT}\right)^2 = \frac{1}{15}\left(\frac{8.2 \cdot 10^{-27}\,\mathrm{C\,m} \cdot 10^5\,\mathrm{V/m}}{1.38 \cdot 10^{-23}\,\mathrm{J\,K^{-1}} \cdot 300\,\mathrm{K}}\right)^2 = 0.0026 \tag{A.54}$$

$$P_\mu = \frac{\varrho \mu^2 E}{3kT} = \varepsilon_0 \Delta \varepsilon E \tag{A.55}$$

dipole contribution $\Delta \varepsilon$ to relative dielectric permittivity ε:

$$\Delta \varepsilon = \frac{\varrho \mu^2}{3\varepsilon_0 kT} \tag{A.56}$$

Electric Field of a Ferroelectric Layer

Electric field at a distance z above the center of a thin charged disk of radius R and charge density σ

$$\mathbf{E} = E_z \hat{\mathbf{z}} = \frac{z}{4\pi\varepsilon_0} \int_0^R \frac{2\pi\sigma\rho\,\mathrm{d}\rho}{r^3}\hat{\mathbf{z}} \tag{A.57}$$

$$r = \left(\rho^2 + z^2\right)^{1/2} \Rightarrow \frac{\mathrm{d}r}{\mathrm{d}\rho} = \rho\left(\rho^2 + z^2\right)^{-1/2} = \frac{\rho}{r} \Rightarrow \rho\mathrm{d}\rho = r\mathrm{d}r \quad (A.58)$$

therefor

$$E_z = \frac{\sigma z}{2\varepsilon_0}\int_0^R \frac{\rho\mathrm{d}\rho}{r^3} = \frac{\sigma z}{2\varepsilon_0}\int_z^{(R^2+z^2)^{1/2}} \frac{\mathrm{d}r}{r^2} = \frac{\sigma z}{2\varepsilon_0}\left[\frac{1}{|z|} - \frac{1}{(R^2+z^2)^{1/2}}\right]$$

$$= \frac{\sigma}{2\varepsilon_0}\left[\mathrm{sgn}\,(z) - \frac{z/R}{\left[1 + (z/R)^2\right]^{1/2}}\right] \quad (A.59)$$

electric field of two charged disks with opposite charge densities at $z = +d/2$ and $z = -d/2$:

$$E_z = \frac{\sigma}{2\varepsilon_0}\left[\mathrm{sgn}\,(z - d/2) - \frac{(z - d/2)/R}{\left[1 + ((z - d/2)/R)^2\right]^{1/2}}\right.$$

$$\left. - \mathrm{sgn}\,(z + d/2) + \frac{(z + d/2)/R}{\left[1 + ((z + d/2)/R)^2\right]^{1/2}}\right] \quad (A.60)$$

Appendix B

Surrface Tension σ & Equilibrium Volume Fraction φ_e

Surface Tension For One-Component Nuclei

With the following calculations based on Mersmann's work [56] for a spherical molecule we show the dependency of interfacial tension σ on the logarithm of equilibrium volume fraction φ_e.
Total surface of a spherical molecule of diameter $d_m = 2r_m$ is:

$$a_{m,tot} = 4\pi r_m^2 = \pi d_m^2, \qquad (B.1)$$

and its volume is:

$$v = \frac{4}{3}\pi r_m^3 = \frac{1}{6}\pi d_m^3, \qquad (B.2)$$

but at the interface a part of molecular surface is not exposed to the liquid phase and does not contribute to the whole surface. Mersmann assumed that it is half of the total molecular surface [56]:

$$a_m = \frac{1}{2}a_{m,tot} = \frac{\pi}{2}d_m^2 = \frac{\pi}{2}\left(\frac{6v}{\pi}\right)^{2/3} = \frac{1}{2}6^{2/3}\pi^{1/3}v^{2/3} = \frac{v^{2/3}}{0.413567}, \qquad (B.3)$$

then

$$\frac{2}{6^{2/3}\pi^{1/3}} = \frac{v^{2/3}}{a_m} = 0.413567, \qquad (B.4)$$

the volume is assumed to be equal to the volume contribution to the solid phase with the molar concentration c^S:

$$v = \frac{1}{c^S N_A} = \frac{M}{\varrho^S N_A}, \qquad (B.5)$$

where M is the molecular weight, ϱ^S is the mass fraction of solute in the solid phase and N_A is the Avogadro constant. Based on Mersmann findings, the surface tension for one-component crystal is define as:

$$\sigma = 0.413567\, kT \left(c^S N_A\right)^{2/3} \ln\left(\frac{c^S}{c^L}\right)$$

$$= \frac{2}{6^{2/3}\pi^{1/3}} kT \left(\frac{\varrho^S N_A}{M}\right)^{2/3} \ln\left(\frac{1}{\varphi_e}\right) = \frac{2}{6^{2/3}\pi^{1/3}} kT v^{-2/3} \ln\left(\frac{1}{\varphi_e}\right)$$

$$= \frac{2kT}{a_{m,tot}} \ln\left(\frac{1}{\varphi_e}\right) = -\frac{2kT}{a_{m,tot}} \ln\varphi_e, \qquad (B.6)$$

here c^L is the molar concentration of solute in the liquid phase and just in vicinity of the interface which is equal to equilibrium volume fraction φ_e of solute in the solution [51]. In the B.6 it is assumed that molar concentration of solute in the solid phase c^S is 100 %. Now one can write:

$$\left(\frac{\sigma}{kT}\right)^3 v^2 = \left[\frac{2}{6^{2/3}\pi^{1/3}} \ln\left(\frac{1}{\varphi_e}\right)\right]^3 = -\frac{2^3}{6^2 \pi} (\ln\varphi_e)^3 = -\frac{8}{36\pi} (\ln\varphi_e)^3$$

$$= -\frac{2}{9\pi} (\ln\varphi_e)^3, \qquad (B.7)$$

with $\beta = 16\pi/3$ for spherical nuclei:

$$\beta\left(\frac{\sigma}{kT}\right)^3 v^2 = -\frac{32}{27} (\ln\varphi_e)^3, \qquad (B.8)$$

and with $Q = 3/5$ for diffusion controlled growth:

$$Q\beta \left(\frac{\sigma}{kT}\right)^3 v^2 = -\frac{32}{45} (\ln\varphi_e)^3 = -0.7111 (\ln\varphi_e)^3. \qquad (B.9)$$

and

$$\sigma = -\frac{kT}{a_m} \ln\varphi_e. \qquad (B.10)$$

Appendix B

Surface Tension For Two-Component Nuclei

If we assume that solute concentration in the solid phase is not 100 % but some solvent molecules present in that phase as well, then volume fraction of the solute in the nuclei $c^S = \varphi'_e$ is $\varphi'_e \leq 1$. Therefore we need to rewrite Mersmann's equation (B.6) with the new assumption:

$$\sigma = 0.413567 kT \left(c^S N_A\right)^{2/3} \ln\left(\frac{c^S}{c^L}\right)$$

$$= \frac{2}{6^{2/3}\pi^{1/3}} k_B T \left(\frac{\varrho^S N_A}{M}\right)^{2/3} \ln\left(\frac{\varphi'_e}{\varphi_e}\right) = \frac{2}{6^{2/3}\pi^{1/3}} kT v^{-2/3} \ln\left(\frac{\varphi'_e}{\varphi_e}\right)$$

$$= \frac{2kT}{a_{\mathrm{m,tot}}} \ln\left(\frac{\varphi'_e}{\varphi_e}\right) = \frac{2kT}{a_{\mathrm{m,tot}}} \left(\ln \varphi'_e - \ln \varphi_e\right), \qquad (B.11)$$

and

$$\left(\frac{\sigma}{kT}\right)^3 v^2 = \left[\frac{2}{6^{2/3}\pi^{1/3}} \ln\left(\frac{\varphi'_e}{\varphi_e}\right)\right]^3 = \frac{2^3}{6^2 \pi} \left(\ln \varphi'_e - \ln \varphi_e\right)^3$$

$$= \frac{8}{36\pi} \left(\ln \varphi'_e - \ln \varphi_e\right)^3 = \frac{2}{9\pi} \left(\ln \varphi'_e - \ln \varphi_e\right)^3, \qquad (B.12)$$

with $\beta = 16\pi/3$ for spherical nuclei:

$$\beta \left(\frac{\sigma}{kT}\right)^3 v^2 = \frac{32}{27} \left(\ln \varphi'_e - \ln \varphi_e\right)^3, \qquad (B.13)$$

and with $Q = 3/5$ for diffusion controlled growth:

$$Q\beta \left(\frac{\sigma}{kT}\right)^3 v^2 = -\frac{32}{45} \left(\ln \varphi_e\right)^3 = 0.7111 \left(\ln \varphi'_e - \ln \varphi_e\right)^3. \qquad (B.14)$$

and

$$\sigma = -\frac{kT}{a_m} \ln \frac{\varphi'_e}{\varphi_e} \qquad (B.15)$$

now one can rewrite the critical value in terms of Gibb's free energy. The maximum of $\Delta G(r)$ is:

$$\Delta G^* = \frac{\beta v^2 \sigma^3}{(kT \ln S)^2} = \frac{(\ln \varphi'_e - \ln \varphi_e)^3}{(\ln \varphi_p - \ln \varphi_e)^2} \frac{\beta v^2}{a_m^3} kT, \qquad (B.16)$$

and the minimum of $\Delta G^* (\varphi_e)$ for

$$\frac{\partial}{\partial \ln \varphi_e} \Delta G^* = \frac{\beta v^2}{a_m^3} kT \frac{\partial}{\partial \ln \varphi_e} \left[\frac{(\ln \varphi'_e - \ln \varphi_e)^3}{(\ln \varphi_p - \ln \varphi_e)^2} \right] = 0, \qquad (B.17)$$

$$\frac{\partial}{\partial \ln \varphi_e} \left[\frac{(\ln \varphi'_e - \ln \varphi_e)^3}{(\ln \varphi - \ln \varphi_e)^2} \right] = \frac{2 (\ln \varphi'_e - \ln \varphi_e)^3}{(\ln \varphi - \ln \varphi_e)^3} - \frac{3 (\ln \varphi'_e - \ln \varphi_e)^2}{(\ln \varphi - \ln \varphi_e)^2} = 0, \quad (B.18)$$

which is solved by

$$\ln \varphi_e = 3 \ln \varphi_p - 2 \ln \varphi_e = \ln \varphi_p^3 - \ln \varphi_e'^2 = \ln \frac{\varphi_p^3}{\varphi_e'^2}, \qquad (B.19)$$

therefore:

$$\varphi_e = \frac{\varphi_p^3}{\varphi_e'^2}, \qquad (B.20)$$

this means that there is a maximum in J and N with respect to φ_e at $\varphi_e = \varphi^3/\varphi_e'^2$.

Acknowledgments

First, I would like to thank my supervisor, Prof. Dr. Günter Reiter for his guidance and support appreciation, and encouragement for my research work at University of Freiburg in the past four years. As an educator, he tried to raise his students as independent researchers and scientists. He has been tremendous help in my dissertation writing and presentation.

I am forever grateful to Dr. Ioan Botiz, Dr. Werner Stille and Dr. Renate Reiter, my colleagues in the experimental polymer physics group, for advising me and taking their valuable time and providing great input into this work.

I am thankful to Dr. Ralf thomann from Freiburger Materialforschungszentrum (FMF) for the electron diffraction measurements, Dr. Harald Scherer from Institute für Anorganische und Analytische Chemie at University of Freiburg for the NMR measurements and Mrs. Barbara Heck, my colleague in the experimental polymer physics group for the X-ray measurements.

I am grateful to Prof. Gert Strobl from our group, Prof. Achim Kittle from University of Oldenburg and Prof. Bernard Lotz from University of Strasbourg, for fruitful discussions and valuable supports.

I also would like to thank other friends and group members, past and present, at University of Freiburg who helped me get through all these years and made them enjoyable for me: Dr. Khosrow Rahimi, Dr. Roozbeh Shokri, Reza Kakavandi, Sejedehsadat Motamen, Dr. Mithun Chowdhury, Dr. Nandita Basu, Dr. Alexander Schmatulla, Dr. Adam Raegen, Andreas Peukert, Shu Zhu, Hui Zhang, Bin Zhang, Rabih Ajib, Stefanie Dold and Silvia Siegenführ.

I would like to express my deepest appreciation to my family for their continuous support in all my life.

I would like to mention: Nafiseh my loving loyal devoted wife, Arad and Artin our sweet lovely twin boys, thank you for your love and unconditional support, without which this would not have been possible. This is for you.

Finally, I acknowledge University of Freiburg and Deutsche Forschungsgemeinschaft (DFG) for supportting this work.

Kaiwan Jahanshahi
Freiburg, August 2013

yes
i want morebooks!

Buy your books fast and straightforward online - at one of world's fastest growing online book stores! Environmentally sound due to Print-on-Demand technologies.

Buy your books online at
www.get-morebooks.com

Kaufen Sie Ihre Bücher schnell und unkompliziert online – auf einer der am schnellsten wachsenden Buchhandelsplattformen weltweit! Dank Print-On-Demand umwelt- und ressourcenschonend produziert.

Bücher schneller online kaufen
www.morebooks.de

 VDM Verlagsservicegesellschaft mbH
Heinrich-Böcking-Str. 6-8　　Telefon: +49 681 3720 174　　info@vdm-vsg.de
D - 66121 Saarbrücken　　　Telefax: +49 681 3720 1749　　www.vdm-vsg.de

Printed by Books on Demand GmbH, Norderstedt / Germany